"This book makes a convincing case for why the East Asia and Pacific region needs to pursue more cooperative approaches to climate action and why the more powerful countries in the region should pursue new approaches to climate leadership."

Miranda Schreurs, *Professor of Environment and Climate Policy, Technical University of Munich, Germany*

"East Asia and the Pacific are on the front lines of climate change, while regional responses will be crucial for global efforts to address the climate crisis. This important new book examines these dynamics, pointing to challenges and pathways for the emergence of a distinct form of regional climate leadership."

Matt McDonald, *Professor of Political Science and International Studies, University of Queensland, Australia*

Regional Climate Leadership in East Asia and the Pacific

This book defines regional climate leadership in East Asia and the Pacific as a novel addition to the climate leadership theory. It develops criteria for measuring such leadership on a country basis and uses these for assessing the efforts of developed, lesser-developed and developing countries within these regions.

The book suggests that regional climate leadership consists of leading domestic actions, leading actions within the region, leading actions that are regionally coordinated and leading actions on a differentiated basis between countries with greater and lesser capacity, and with neighbourly intent. The book is policy and climate solutions-focused and identifies opportunities for lesson learning and policy transfer for more effective mitigation and adaptation. These solutions take into account the widely varying and complex geographical, political and institutional circumstances of the region.

It is intended for a broad readership of climate policy actors, including policy professionals, academics, non-government researchers and all who are looking for climate leadership solutions to the problems of accelerating climate impact in East Asia and the Pacific.

Kate Crowley, Adjunct Associate Professor of Public and Environmental Policy at the University of Tasmania, has widely published on environmental politics and climate policy. She is the recipient of the Vice-Chancellor of the University of Tasmania's award for Outstanding Community Engagement on climate change and sustainability.

Akihiro Nakamura has been Associate at Meiji University, the University of Tasmania, and the Institute of Social Sciences of the Central Region, Viet nam. He is also a climate policy researcher/analyst at the Research Institute of Innovative Technology for the Earth, Japan, and IOM Law, Norway.

Routledge Advances in Management and Business Studies

How to Manage and Survive during a Global Crisis
Lessons for Managers from the COVID-19 Pandemic
Piyush Sharma and Tak Yan Leung

Understanding Procrastination at Work
Individual and Workplace Perspectives
Beata Bajcar

Entrepreneurship in the BRICS
Economic Development and Growth in the Post-Pandemic World
Edited by Ndivhuho Tshikovhi, Fulufhelo Netswera and Ravinder Rena

Start-up Strategy and Entrepreneurial Development
Evolving Business Models and Economic Trends
Edited by Iwona Skalna and Aleksandra Gąsior

Enterprise Risk Management
Technical, Cognitive, and Social Perspectives
Mirna Jabbour and Jason Crawford

Navigating the Ecological Transition
A Business School Perspective
Edited by Hugues Bouthinon-Dumas, Arijit Chatterjee and Bernard Leca

Regional Climate Leadership in East Asia and the Pacific
Kate Crowley and Akihiro Nakamura

For more information about this series, please visit: www.routledge.com/Routledge-Advances-in-Management-and-Business-Studies/book-series/SE0305

Regional Climate Leadership in East Asia and the Pacific

Kate Crowley and Akihiro Nakamura

Routledge
Taylor & Francis Group
LONDON AND NEW YORK

First published 2025
by Routledge
4 Park Square, Milton Park, Abingdon, Oxon, OX14 4RN

and by Routledge
605 Third Avenue, New York, NY 10158

Routledge is an imprint of the Taylor & Francis Group, an informa business

British Library Cataloguing-in-Publication Data
A catalogue record for this book is available from the British Library

ISBN: 978-0-367-77235-2 (hbk)
ISBN: 978-0-367-77233-8 (pbk)
ISBN: 978-1-003-17038-9 (ebk)

DOI: 10.4324/9781003170389

Typeset in ITC Galliard Pro
by KnowledgeWorks Global Ltd.

Contents

Figures

Tables

Foreword

The policy choices made by countries in the East Asia and Pacific region will have a decisive impact on the global community's chances of reducing greenhouse gas emissions and adapting to the risks associated with climate change. The region is a mix of some of the strongest economies alongside some of its weakest and most vulnerable. It is politically, culturally and religiously diverse perhaps explaining why the sense of being a regional community remains weak despite the strong economic interdependencies that exist among many states in the region. This matters for climate change. Domestic policies alone will not suffice to address the causes and consequences of climate change.

If the larger and economically more developed states of the region do not do more to abet their greenhouse gas emissions, not only domestically but also regionally (and globally), then the smaller and more vulnerable states of East Asia and the Pacific will face even more devastation from climate-related extreme weather events, including heat extremes, severe rains, hurricanes and rising sea levels. The Climate Action Tracker, which assesses major countries' climate mitigation performance, finds that all of the major players in the region are doing too little to meet the goals set by the Paris Agreement. While China, Japan, South Korea, Taiwan, Australia and New Zealand have established Nationally Determined Contributions and climate neutrality goals, domestic policy implementation lags and regional climate leadership vision is weak or missing.

The member states of the United Nations agreed that combatting climate change will require common but differentiated measures and that developed states must assist developing states adapt to climate change. This book makes a convincing case for why the East Asia and Pacific region needs to pursue more cooperative approaches to climate action and why the more powerful countries in the region should pursue new approaches to climate leadership. Of course, domestic climate action in China, the world's largest greenhouse gas emitter, for example, matters tremendously. China has shifted course in some areas and is now the world leader in renewable energy capacity expansion and in the development of electric automobiles. Yet it has also invested heavily in the expansion of coal-fired power plants abroad and in resource extraction from across East Asia and the Pacific. Janus-faced policies – which clean up actions domestically but shift emissions or environmental impacts abroad – do little to mitigate against climate change or biodiversity loss. Ambitious climate policies are only meaningful if they are implemented in a timely fashion.

Failing to help the most vulnerable states of East Asia and the Pacific to adapt to climate change will only mean higher disaster-related costs and greater human suffering. Crowley and Nakamura rightly point out that the climate crisis is already upon us. The time to face up to this reality is now, and the economically powerful states of East Asia and the Pacific should consider what they can do to lead globally in the fight against climate change by first becoming regional climate leaders. This will involve considering the regional implications of domestic policy actions, developing regionally focused policy instruments and programs and aiding the smaller economies in the region to adapt to climate change. It will also mean aiding rapidly growing economies like Indonesia and the Philippines, which are growing both in terms of their populations and their emissions to transition more rapidly to renewable energies and energy efficiency.

This book is an important addition to the literature on climate policy in East Asia and the Pacific region. It takes a refreshing new look at the potential of the region to lead on climate change and makes a powerful call for the bigger countries of the region to grasp this leadership role. Waiting for others to act may mean that for many states in the region, it will be too late. Strong domestic climate action linked to intensified regional mitigation and adaptation efforts are needed, and they are needed now.

Miranda Schreurs, Professor of Environment and Climate Policy,
Technical University of Munich

Abbreviations

AAP-JRCC	ASEAN Action Plan on Joint Response to Climate Change
ADB	Asian Development Bank
AMS	ASEAN member states
AOSIS	Alliance of Small Island States
AP6	ASEAN Plus Six (+6)
APAP	Asia-Pacific Action Plan 2021–2024 for the implementation of the Sendai Framework for disaster risk reduction 2015–2030
APEC	Asia-Pacific Economic Cooperation
APT	ASEAN Plus Three (+3)
ARF	ASEAN Regional Forum
ASEAN	Association of Southeast Asian Nations
ASPI	Asia Society Policy Institute
BEAPA	Bureau of East Asian and Pacific Affairs
BRI	China's Belt and Road global investment initiative
CAT	Climate action tracker
CBAM	Carbon border adjustment measure
CBDR	Common but differentiated responsibility
CCAP	Climate change adaptation policy
CCPI	Climate Change Performance Index
CCS	Carbon capture and storage
COP	Conference of Parties
CPM	Carbon pricing mechanism
CREA	Centre for Research on Energy and Clean Air
CRI	Germanwatch Global Climate Risk Index
CSIRO	Commonwealth Scientific and Industrial Research Organization
CT	Carbon tax
DELSA	Disaster Emergency Logistic System for ASEAN
DFAT	Department of Foreign Affairs and Trade (Australia)
DRR	Disaster risk reduction
EA&P	East Asia and the Pacific
EAF	East Asia Forum
EAS	East Asia Summit

EIA	U.S. Energy Information Administration
ECFR	European Council for Foreign Relations
ESCAP	United Nations Economic and Social Commission for Asia and the Pacific
ESCAPCDRR	ESCAP Committee on Disaster Risk Reduction
ETS	Emissions trading scheme
EU	European Union
FFN-PTI	Fossil Fuel Non-Proliferation Treaty Initiative
FRDP	Framework for Resilient Development in the Pacific
GCF	Green Climate Fund
GEF	Global Environment Facility
GHGs	Greenhouse gas emissions
ICAP	International Carbon Action Partnership
IEA	International Energy Agency
IGES	Institute for Global Environmental Strategies
ILO	International Labour Organisation
IMF	International Monetary Fund
INDC	Intended Nationally Determined Contribution
IOM	International Organisation for Migration
IPCC	United Nations Intergovernmental Panel on Climate Change
IRENA	International Renewable Energy Agency
ISM on DR	Inter-Sessional Meetings on Disaster Relief
ITMOs	Internationally Transferrable Mitigation Outcomes
JAIF	Japan-ASEAN Integration Fund
JETPs	Just Energy Transition Partnerships
KETS	Korean Emissions Trading Scheme
LDCF	Least Developed Countries Fund
LDCs	Least developed countries
LIC	Lower-income countries
LSE	London School of Economics and Political Science
LULUCF	Land Use, Land Use Change and Forestry
ND-GAIN	Notre Dame Global Climate Adaptation Initiative
NDCs	Nationally determined contributions
NGOs	Non-governmental organisations
ODA	Official development assistance
OECD	Organisation for Economic Cooperation and Development
OHCHR	Office of the High Commissioner for Human Rights
PAFTAD	Pacific Trade and Development Conference
PBEC	Pacific Basin Economic Council
PCCMHS	UN Pacific Climate Change Migration and Human Security Program
PECC	Pacific Economic Cooperation Council
PICs	Pacific Island Countries
PIDF	Pacific Island Development Forum
PIF	Pacific Island Forum

PIFS	Pacific Islands Forum Secretariat
PRC	Pew Research Center
PwC	Price Waterhouse Cooper Global
RCT	Regional carbon trading
REDD	Reducing emissions from deforestation and forest degradation
RERT	Regional emissions reduction target
RGGI	Regional Greenhouse Gas Initiative
ROK	Republic of Korea
SAR	IPCC's Sixth Assessment Report (SAR1, SAR2, SAR3, etc.)
SCCF	Special Climate Change Fund
SCN	ASEAN Strategy for Carbon Neutrality
SDGs	Sustainable Development Goals
SIDS	Small Island Developing States
SPC	South Pacific Community
SPREP	Secretariat of the Pacific Regional Environment Programme
SSCF	South-South Cooperation Fund (China)
TEMM	Tripartite Environment Ministers Meeting
UN	United Nations
UNCCS	United Nations Climate Change Secretariat
UNDP	United Nations Development Programme
UNESCAP	United Nations Economic and Social Commission for Asia and the Pacific
UNFCCC	United Nations Framework Convention on Climate Change
UNFCCC/RCC	The Regional Collaboration Center for Asia-Pacific of the Secretariat of the United Nations Framework Convention on Climate Change
UNIDO	United Nations Industrial Development Organization
UNISDR	United Nations International Strategy for Disaster Reduction
UNDRR	United Nations Office for Disaster Risk Reduction
US	United States
USP	University of the South Pacific
WEF	World Economic Forum
WHO	World Health Organization
WMO	World Meteorological Organisation
WPR	World Population Review
WRI	World Resources Institute
WTO	World Trade Organisation

'Forget the poetic flap of a butterfly's wings in Beijing causing rain in Central Park. Climate issues in Asia-Pacific are measured in superlatives. The world's biggest population. Two of the three largest carbon dioxide-emitting countries and the largest share of emissions globally. The most exposed to extreme weather events. Some of the smallest and most vulnerable countries. Also, the fastest-growing part of the global economy and many of the leaders in green technology.

What Asia does to fight global warming will be literally felt across the whole planet.'

(Gaspar and Yong Rhee, 2021)

1 Introduction

Introduction and Background

The East Asia and Pacific region (EA&P) is extremely vulnerable to climate change because of overpopulation, extreme poverty, the greenhouse gas (GHG) emissions burden of rapid economic development and high baseline levels of flooding, drought and extreme temperatures (World Bank, 2022a). It is the most disaster-stricken region in the world, with a lack of capacity to build resilience in terms of preparedness and response, and is now experiencing climate change impacts that are outstripping its adaptation efforts. The region's capacity to respond is hampered not only by the predominantly lesser-developed and developing status of many EA&P countries but also by the complex and competing regional architectures and organisations within it. None of the EA&P countries perform well on the Climate Change Performance Index (CCPI, 2024). Thailand is moderately ranked (medium), for example, China and Viet Nam are lowly ranked and Malaysia and the Democratic People's Republic of Korea (South Korea) are very lowly ranked. However, developed countries in the EA&P, from which leadership could be expected, rank poorly on the CCPI[1] with New Zealand (low), Australia (low) and Japan (very low) ranked, respectively, at 34th, 50th and 58th out of 63 countries.

Against the backdrop of this universally poor climate policy performance, we argue in this book for regional climate leadership that facilitates urgent climate action in EA&P, but not simply in terms of national performance. Whilst domestic emissions reduction of nation states is a bedrock performance indicator of climate action, we argue for a broader notion of regional climate leadership that is suited to the challenges facing EA&P. We suggest that climate leadership by a nation state within EA&P would embrace action with both domestic and neighbourly, regional intent, both within and beyond nation state borders, that outperforms other countries certainly, but in a polycentric context whereby domestic action also addresses regional challenges. Our normative criteria for measuring regional climate leadership, introduced in Chapter 3, therefore, include domestic climate policy credibility and effectiveness, but also responsible and coherent national actions

DOI: 10.4324/9781003170389-1

within the region, and action to build regional capacity that recognises common but differentiated country responsibility (Table 3.3). Our approach is policy and outcome focused, seeking to draw lessons, to promote capacity and to suggest solutions, as much as to identify and contrast climate policy performance.

The means of achieving effective climate policy governance and action to reduce GHG emissions has been guided for the last few decades by international conventions, protocols and agreements. The *United Nations Framework Convention on Climate Change* (UNFCCC), a convention that was initiated by the 1992 *Rio Earth Summit* and adopted in 1994, is the parent treaty of both the *Kyoto Protocol* and the subsequent *Paris Agreement.* Despite being adopted in 1997 by 192 UNFCCC parties, the *Kyoto Protocol* was not ratified until 2005. For a variety of reasons, not least of which were the economic and political obstacles to action at the level of the nation state, global GHG emissions had risen steeply by the conclusion in 2020 of the second *Kyoto Protocol* commitment period, known as the Doha amendment. Earlier, in December 2015, 196 nation states had adopted a successor agreement to the *Kyoto Protocol*, namely the *Paris Agreement*, which is a legally binding international treaty to limit global warming to well below 2°C and preferably to 1.5°C by 2050, compared to pre-industrial levels (EPRS, 2015). Together these instruments have aimed to stabilise GHG emission concentrations to prevent dangerous, anthropogenic (human induced) interference with the climate system as determined by regular advice from the United Nations Intergovernmental Panel on Climate Change (IPCC).

Despite United Nations conventions, protocols and agreements introducing a context and process for accelerated global GHG reduction, by 2022 global GHG emissions had increased by an alarming 50% on 1990 levels to almost 50 billion metric tons of carbon dioxide equivalent ($GtCO_2e$) (Statista, 2023a). Recent reports by the IPCC have unequivocally highlighted the need for the global community to act urgently now to limit global warming as close to 1.5°C above pre-industrial levels as possible if the catastrophic impacts of climate change are to be averted. A rapid shift across all states, developed, lesser-developed and developing, and indeed across all sectors, is needed imminently to accelerate the transition to a low carbon economy and a less carbon-endangered world. Whilst the *Kyoto Protocol* had embraced prior action by developed countries, the failure of these countries to effect sufficient emissions reductions has seen the less affluent, lesser-developed and developing countries drawn into the emissions reduction effort under the *Paris Agreement.* However, global action is falling dangerously short.

In terms of providing climate advice on how to proceed, the IPCC was in its sixth assessment cycle in 2022, during which time a report from each of its three working groups (physical science; impacts, adaptation and vulnerability; and mitigation) was delivered.

In releasing the first instalment of the IPCC's Sixth Assessment Report (SAR), António Guterres, the Secretary General of the United Nations, warned that it is a 'code red for humanity' with irrefutable evidence of human influence upon climate change. Guterres also warned that the 'alarm bells are deafening', and 'the evidence irrefutable', namely that 'greenhouse-gas emissions from fossil-fuel burning and deforestation are choking our planet and putting billions of people at immediate risk'. Global warming is no longer a sufficient descriptor when 'global heating is affecting every region on Earth, with many of the changes becoming irreversible'. He highlighted the SAR findings that only ambitious, urgent and decisive action by nation states to deeply cut GHG emissions will suffice to limit the global temperature rise to 1.5°C (UN, 2021).

In launching the second instalment of the IPCC's Sixth Assessment Reports (SAR2), on impacts, adaptation and vulnerability, Guterres lambasted the catalogue of broken promises and empty pledges by nation state leaders that have the planet on a fast track to disaster, with:

> Major cities under water. Unprecedented heatwaves. Terrifying storms. Widespread water shortages. The extinction of a million species of plants and animals. This is not fiction or exaggeration. It is what science tells us will result from our current energy policies. We are on a pathway to global warming of more than double the 1.5°C limit agreed in Paris. Some Government and business leaders are saying one thing but doing another. Simply put, they are lying. And the results will be catastrophic. This is a climate emergency.
>
> (UN, 2022a)

The last instalment of the Sixth Assessment Report (SAR3) on mitigation potential aims to accelerate GHG emissions reduction by promoting effective actions that are already in place but that urgently need to be scaled up globally, and that need to proliferate across the EA&P. Rapid uptake of solar and wind energy and batteries, for example, has seen their cost plummet by 85% over the last decade proving that 'we have the tools and the know how to limit [global] warming'. This latest report acknowledges the proliferation of 'policies and laws [that] have enhanced energy efficiency, reduced rates of deforestation and accelerated the deployment of renewable energy'. But it is 'now or never' if global warming is to be limited to 1.5°C by 2050 with global GHG emissions needing to peak by 2025 and reduce to 43% by 2030. Methane, incidentally, must be reduced as well by 30% by 2030 (IPCC, 2022a). The *Paris Agreement* includes the expectation that developed, lesser-developed and developing countries draw upon IPCC reports in submitting their non-binding nationally determined contributions (NDCs), which detail their domestic policies, measures and plans for global emissions reduction that would ideally see global warming limited as close to 1.5°C as possible by 2050.

It is within this context of the clear need for very urgent climate policy action, to build capacity for emissions reduction action and to transition to a low carbon future, that we explore the potential for regional climate leadership in EA&P. The United Nations Economic and Social Commission for Asia and the Pacific recognises that mitigation efforts by Asia-Pacific member states are not aligned with the goals of the Paris Agreement. It has called for more robust and accurate emissions reduction scenarios for the next ten years to be developed, as part of the 2024 NDC update cycle, detailing immediate, substantial and sustained steps by member states to reduce their GHG emissions (UNESCAP, 2023a). In the absence of an EU style of formalised and regulatory regional governance, and the fractured and complex regional context of EA&P, we consider regional leadership as domestic efforts that extend to the region. In this chapter, we introduce and define the EA&P region, the challenges and impacts of climate change within it and the need for leadership to be defined and practised within a regional context to accelerate the climate action that is so urgently required.

East Asia and the Pacific (EA&P)

The IPCC defines several Asian regions: West Asia, Central Asia, South Asia, North Asia, East Asia and Southeast Asia, which stretch from the Middle East to Papua and New Guinea from west to east, and from the Arctic to Indonesia from north to south (IPCC, 2014). It is East and Southeast Asia, and the Asia Pacific, that interests us, for enabling us to focus on the arc of nation states from East Asia, including China, eastward across the Pacific, to the United States and for capturing the myriad of sub-regions in between. Broadly defined, the region of interest to us aligns with the EA&P as defined by the US Bureau of East Asian and Pacific Affairs (2017), which includes China but excludes the United States. This definition includes a broad range of lesser-developed countries, developing countries, Pacific Island countries and the developed countries of Australia, Japan, New Zealand, South Korea and potentially even ROC Taiwan[2] (Taiwan), which is an autonomous region, from which regional climate leadership could be expected; noting, however, that landlocked Mongolia is excluded from our research (Table 1.1).

The Bureau of East Asian and Pacific Affairs (Bureau) inclusion of developed countries in their definition provides the opportunity that we have previously identified (Crowley & Nakamura, 2017) for examining developed countries as catalysts and supporters of lesser-developed and developing countries in terms of achieving rapidly accelerating climate action within the EA&P region. The EA&P region is defined by the Bureau in terms of the United States' interests, for the purposes of promoting regional engagement, peace, mutually beneficial relations, freedom, denuclearisation and, specifically, a 'results-oriented'

Table 1.1 East Asia and the Pacific – the region defined

Region	*Source*	*Countries*
East Asia and Pacific (EA&P)	Bureau of East Asian and Pacific Affairs (BEAPA)	Australia, Brunei, Myanmar, Cambodia, China (inc. Hong Kong Special Administrative Region and Macau Special Administrative Region), Timor-Leste, Fiji, Indonesia, Japan, Kiribati, Lao PDR, Malaysia, Marshall Islands, Micronesia Fed. Sts., Mongolia (excluded), Nauru, New Zealand, North Korea, Palau, Papua New Guinea, Philippines, Samoa, Singapore, Solomon Islands, South Korea, Taiwan, Thailand, Tonga, Tuvalu, Vanuatu, Viet Nam
East Asia and the Pacific	World Bank	American Samoa, Australia, Brunei Darussalam, Cambodia, China, Fiji, French Polynesia, Guam, Hong Kong SAR China, Indonesia, Japan, Kiribati, South Korea, Korea, Rep., Lao PDR, Macao SAR China, Malaysia, Marshall Islands, Micronesia, Fed. Sts., Mongolia, Myanmar, Naru, New Caledonia, New Zealand, North Mariana Islands, Palau, Papua New Guinea, Philippines, Samoa, Singapore, Solomon Islands, Thailand, Timor-Leste, Tonga, Tuvalu, Vanuatu, Viet Nam
East Asia	IPCC	People's Republic of China, Hong Kong Special Administrative Region, Macao Special Administrative Region, Japan, North Korea, South Korea, Taiwan
Southeast Asia	IPCC	Brunei, Indonesia, Lao People's Democratic Republic, Malaysia, Myanmar, Papua New Guinea, Philippines, People's Republic of Cambodia, Singapore, Thailand, Timor-Leste, Viet Nam

Sources: APSCC (2023); BEAPA (2017); IPCC (2014); World Bank (2022a).

relationship with China (BEAPA, 2017). Our interest, however, is in exploring regional climate leadership with neighbourly and regional intent, with a deliberate focus beyond the superpowers of China and the United States, although China's influence, intent and actions are considered in Chapter 7 in terms of its leading and lagging climate policy efforts. The EA&P region must build its climate policy capacity, and working synergistically and cooperatively with leadership from countries with the greatest capacity will accelerate its own capacity building efforts.

In Chapter 2 we review the geographic and political complexity of the EA&P region as an additional challenge to regional climate action, but also as an opportunity to implement and potentially join up bilateral, polycentric and multi-levelled action in the interests of achieving rapid change. It is worth noting that ASEAN, for example, which is discussed in Chapter 2, is a narrowly defined lesser-developed and developing nation grouping within the broader EA&P that is nevertheless crucial to understanding progress both on climate action and any potential shift towards carbon trading within the region, for example. We are nevertheless predominantly interested in EA&P for its extended geographical reach between China and the United States, and its inclusion of developed countries, but also for the capacity, and policy leverage, that it offers for globally significant climate action. An extensive network of integrated developed, lesser-developed and developing country climate actions implemented within the region, and at times facilitated by regional arrangements and institutions, could promote greater emissions reduction ambition, ideally guided by a regional emissions reduction target (RERT), and help to build resilience against climate impacts in a manner that aligns with the *Paris Agreement.* Domestic climate achievements of all the EA&P countries could thrive in such a context.

In Table 1.1, we have detailed various defined regions within the vicinity of EA&P beginning with the broadest defined region which we have adopted for our analysis. In Table 1.2, we profile this region in terms of its population, gross domestic product, per capita emissions, % of global emissions and emissions change from 1990 to 2019 of EA&P countries. In 2022, these countries accounted for 40.53% of global GHG emissions, with China responsible for the bulk of those (29.16%) but with Japan (2.20%), South Korea (1.35%), Indonesia (2.31%) and Australia (1.06%) otherwise notable regional contributors (Table 1.2). In 2020, the per capita emissions of the EA&P countries present a starkly differing profile of major emitters (excluding Palau and Brunei) that includes: Australia (21.98), South Korea (14.01), Taiwan (12.86), Singapore (11.67), Japan (9.41), China (10.95), Malaysia (10.50) and New Zealand (16.83) (Table 1.2; Crippa et al., 2023). The countries that appear most economically capable, in terms of their GDP, of reducing their GHG emissions and of assisting other countries in the region are Australia, China, Indonesia, Japan, Malaysia, New Zealand, South Korea, Taiwan and Thailand. This selection of developed, lesser-developed countries and developing is therefore the focus of a more detailed analysis in Chapter 4.

The EA&P region is, without doubt, a vast and challenging region in which to promote climate action, with its differing geographies, resources, economies and fossil fuel dependencies, its vast reach, diverse sets of politics and cultures and a plethora of regional organisations, summits and frameworks. This is in part why we argue that bilateral, polycentric and multi-levelled actions, and a supportive regional context, are required to achieve rapid emissions reduction and change. But the region is united, not only by the daunting scale of its combined, accelerating GHG emissions but also in its experience of the accelerating, destructive impacts of climate change, with extreme weather already having 'killed thousands, displaced millions and cost billions in 2020', which is in one year alone (WMO, 2023).

Table 1.2 EA&P population, GDP per capita, per capita emissions, % global emissions and emission trajectories

Country	*Population*	*GDP per capita*	*2022 Per capita emissions*	*% of Global emissions*	*1990 Emissions*	*2020 Emissions*	*1990–2020 % change*
Sources	CIA (2024)	CIA (2024)	EU Commission – Crippa et al. 2023	EU Commission – Crippa et al. 2023	UNFCCC [kt CO_{2e}] UNFCC (2022)	UNFCCC [kt CO_{2e}] UNFCC (2022)	UNFCCC (2022)
+Australia	26,768,598	$59,500	21.98	1.06	425,624.31 kt CO_{2e}	528,149.46	24.09%
Brunei	491,900	$77,900	32.66	0.03	9,488.8 mt CO_{2e} in 2010	n.a.	n.a.
Cambodia	17,063,669	$5,100	2.91	0.09	118,729.6 mt CO_{2e} in 1990	147,282.2 mt CO_{2e} in 2000	24.05% (1990–2000)
+China	1,416,043,270	$22,100	10.95	29.16	4,057,617.0 mt CO_{2e} in 1994	12,300,200.0 mt CO_{2e} 2014	203.14% (1990–2014)
Fiji	951,611	$13,600	3.19	0.01	1,391.3 mt CO_{2e} in 1994	2710.1 mt CO_{2e} in 2004	94.78% (1994–2004)
+Indonesia	281,562,465	$14,100	4.47	2.31	266,818.4 mt CO_{2e} in 1990	554,333.5 mt CO_{2e} in 2000	107.76% (1990–2000)
+Japan	123,201,945	$46,300	9.41	2.20	1,269,901.03	1,168,094.47	−7.98%
Kiribati	116,545	$3,200	0.87	0.00	28.0 mt CO_{2e} in 1994	170.3 mt CO_{2e} in 2008	508.81% (1994–2008)
Lao PDR	7,953,556	$8,400	5.01	0.07	6,866.6 mt CO_{2e} in 1990	8,898.2 mt CO_{2e} in 2000	29.59% (1990–2000)
+Malaysia	34,564,810	$33,600	10.50	0.66	67,367.9 mt CO_{2e} in 1990	287,740.3 mt CO_{2e} in 2011	327.12% (1990–2011)
Marshall Islands	82,011	$6,800	n.a.	n.a.	122.5 mt CO_{2e} in 2000	169.8 mt CO_{2e} in 2010	38.60% (2000–2010)
Micronesia, Fed. Sts.	99,603	$3,800	n.a.	n.a.	246.0 mt CO_{2e} in 1994	173.9 mt CO_{2e} in 2000	−29.30% (1994–2000)
Myanmar	57,527,139	$5,300	3.04	0.31	33,995.7 mt CO_{2e} in 2000	38,374.9 mt CO_{2e} in 2005	12.88% (2000–2005)
Nauru	9,892	$11,400	n.a.	n.a.	35.9 mt CO_{2e} in 1994	42.2 mt CO_{2e} in 2010	17.51% (1994–2010)
+New Zealand	5,161,211	$48,800	16.83	0.15	65,196.98	76,824.59	18.70%

(Continued)

Table 1.2 (Continued)

Country	*Population*	*GDP per capita*	*2022 Per capita emissions*	*% of Global emissions*	*1990 Emissions*	*2020 Emissions*	*1990–2020 % change*
(North) Rep. Korea	26,298,666	$1,700	3.12	0.15	n.a.	n.a.	n.a.
Palau	21,864	$15,800	61.65	0	104.9 mt CO_{2e} in 1994	346.5 mt CO_{2e} in 2005	230.40% (2000–2005)
Papua New Guinea	10,046,233	$4,200	1.00	0.02	5,012.2 mt CO_{2e} in 1994	10,195.5 mt CO_{2e} in 2000	103.42% (1994–2000)
Philippines	118,277,063	$9,700	2.35	0.49	100,866.6 mt CO_{2e} in 1994	126,878.7 mt CO_{2e} in 2000	25.79% (1994–2000)
Samoa/Am Samoa	208,853/ 43,895	$6,000/ $11,200	2.70	0.00/0.00	560.8 mt CO_{2e} in 1994	n.a.	n.a.
Singapore	6,028,459	$127,500	11.67	0.13	26, 859.1 mt CO_{2e} in 1994	48,333.9 mt CO_{2e} in 2012	79.95% (1994–2012)
Solomon I slands	726,799	$2,700	0.88	0.00	294.4 mt CO_{2e} in 1994	618.6 mt CO_{2e} in 2010	110.14% (1994–2010)
+South Korea	52,081,799	$50,600	14.01	1.35	292,211.2	727,638.4	149.01%
+Taiwan	23,595,274	$47,800	12.86	0.57	n.a.	n.a.	n.a.
+Thailand	69,920,998	$21,100	6.67	0.86	233,990.1 mt CO_{2e} in 1994	318,660.9 mt CO_{2e} in 2013	42.27% (1994–2013)
Timor-L	1,506,909	$4,600	1.72	0.00	1,276.6 mt CO_{2e} in 2010	n.a.	n.a.
Tonga	104,889	$6,600	2.51	0.00	229.2 mt CO_{2e} in 1994	192.2 mt CO_{2e} in 2006	−16.16%
Tuvalu	11,733	$5,200	n.a.	0.00	5.6 mt CO_{2e} in 1994	18.4 mt CO_{2e} in 2014	231.43% (1994–2014)
Vanuatu	318,007	$3,000	1.91	0.00	299.4 mt CO_{2e} in 1994	586.2 mt CO_{2e} in 2000	95.81% (1994–2000)
+Viet Nam	105,758,975	$13,700	4.88	0.91	84,456 mt CO_{2e} in 1994	278,441.9 mt CO_{2e} in 2013	229.70% (1994–2013)

Sources: CIA (2024); Crippa et al. (2023); (2022); Non-Annex 1 countries – https://di.unfccc.int/ghg_profile_non_annex1.

Notes: As defined by BEA&PA (2017); GDP – per capita compares real GDP on a purchasing power parity basis divided by population in July of the same year. GHG total without LULUCF (Land Use, Land Use Change and Forestry) in kt CO_2 equiv. n.a. indicates no data, which can also be interpreted as 0.0% of global emissions. The "+" denotes that this country appears in Chapter 4.

Climate Impacts in East Asia and the Pacific (EA&P)

Climate impacts in EA&P are detailed in the IPCC's 2022 *Impacts, Adaptation and Vulnerability* regional assessment reports in their summary factsheets for Asia and Australasia, with rising temperatures increasing the occurrence of natural disasters – with increased heatwaves, droughts, cyclones, floods, typhoons, biodiversity loss, human diseases and deaths, food and energy insecurity, industrial systems risk and involuntary population displacement. Whilst climate impacts will likely drive involuntary migration across Asia in the future, in Southeast and East Asia, cyclones, floods, and typhoons triggered internal displacement of 9.6 million people in 2019, almost 30% of total global displacements (IPCC, 2022b). As noted above, EA&P is extremely vulnerable to climate change because of overpopulation, extreme poverty, the emissions burden of rapid economic development and high baseline levels of flooding, drought and extreme temperatures. Indeed, it is the most disaster-stricken region in the world and already suffers from a lack of capacity to build resilience in terms of disaster preparedness and response (see Chapter 5) but is now experiencing climate change impacts that are outstripping its adaptation efforts. The World Bank observes further that:

> The region includes 13 of the world's 30 most climate-vulnerable countries. It also bears the brunt of 70% of the world's natural disasters, which have affected more than 1.6 billion people in the region since 2000. Pacific Island countries, where rising sea levels are threatening coastal areas and atoll islands, have been hit hard by climate-induced disasters, including Category Five Tropical Cyclone Harold, which left a path of destruction across Fiji, Solomon Islands, Tonga, and Vanuatu in April 2020. Following that, in December 2020, Category Five Tropical Cyclone Yasa made landfall in Fiji, which caused significant damage and the loss of lives. 2021 has already seen two severe Tropical Cyclones in the Pacific.
>
> (World Bank, 2022a)

The key messages from the most recent *State of the Climate in Asia* report by the World Meteorological Organization (WMO) include that: '(i)n 2023, the mean temperature over Asia was 0.91°C above the 1991–2020 reference period, the second highest on record. Many parts of the region experienced extreme heat events in 2023' (WMO, 2024, p. ii). The region was the most disaster-hit in the world again in 2023, by floods, typhoons, storms, droughts and heatwaves, with Japan also experiencing its hottest summer on record, China experiencing drought, and the overall warming trends of the regions' oceans continuing. Climate related disasters were responsible for deaths and economic costs, with economically vulnerable countries again being disproportionately impacted. The WMO (2024, p. 1) also found that '(t)he year 2023 was the warmest year on record according to six globally averaged datasets, and that the nine years 2015 to 2023 were the nine warmest years on record in all datasets.'

The UNESCAP has described Asia-Pacific sub-regional 'riskscapes' which would exist at 1.5 °C of climate change by 2050, as well as potential pathways for adaptation and resilience (UNESCAP, 2022). Drawing upon the IPCC's SAR, it finds that Asia and the Pacific will be the most impacted by heavy precipitation, followed by agricultural drought, hot temperatures/heatwaves and warming winds with intensifying tropical cyclones. The WMO (2021) identified climate change impacts in Asia in 2020 as killing thousands, displacing millions and costing billions, with havoc wreaked upon populations, infrastructure and ecosystems. Sustainable development goals were suffering, and climate action in some cases was regressing. The cost of past extreme weather events – such as tropical cyclones, floods and droughts – in 2020 in Asia in terms of average annual loss amounted to several hundred billion US dollars: i.e. US $237,971b (in China); US$83,350b (in Japan); US$24,279b (in South Korea); US$12,490b (in Thailand); and US$10,834b (in Viet Nam) (in Viet Nam).

The IPCC has previously drawn attention to the vulnerability of the EA&P region, noting that many of its countries do not have the capacity, technology or resources to mitigate, adapt to or, in some cases, survive the impacts of climate change (IPCC 2014, see chapter 24 'Asia'). This lack of capacity is of interest in governance responses but also to the financial sector, and service providers such as Price Waterhouse Cooper Global (PwC), which is a professional services network that delivers assurance, tax and advisory services including to 85% of the Global Fortune 500 companies. Besides committing to reducing its own GHG emissions to net zero by 2030, and assisting its clients in doing the same, PwC has issued its own 'code red' for the Asia Pacific because the region was decarbonising at only 2.8% in 2022 compared to the 17.2% per annum decarbonisation effort that is required to achieve net zero by 2050 (PwC, 2023a). The region therefore needs to urgently accelerate its decarbonisation efforts *by a factor of at least ten* to avert locking in dangerous climate change but is also projected to unlock a positive boost in stimulatory economic activity from its efforts to do so (World Bank, 2021).

The scale and urgency of the mitigation, adaptation and resilience building challenges across EA&P, which we consider in detail in Chapter 5, are daunting and require combined efforts in response, essentially a cross-sectoral series of polycentric, global, domestic, regional and local actions and policy solutions. Urgent climate action as a principle and practice must be integrated into the fabric of the economy and society on everything, by every country, at all times. We suggest that regional climate leadership, which is EA&P specific in terms of climate mitigation, adaptation, resilience and capacity building actions at the domestic and regional levels, is a significant part of meeting this challenge. Whilst we are primarily interested in leadership within and between nation states in EA&P, regional climate leadership is a flexible, nuanced, multi-levelled notion that could readily capture multi-actor efforts. Further research beyond our scope here could identify and interrogate interlocking public and private actions at the local, national and regional levels and their contribution to global climate change goals.

Leadership and Polycentric Action

Recent IPCC, CCPI and Climate Action Tracker (CAT) reports show that developed, lesser-developed and developing nation states alike are struggling to reduce their GHG emissions sufficient to meet the global community's 2050 target of net zero emissions (of global carbon dioxide). CAT is a collaborative exercise between Climate Analytics and the NewClimate Institute that provides independent scientific analysis of nation state targets, policies and actions against the global goals of the 2015 *Paris Agreement.* It covers the biggest nation state emitters as well as a representative sample of smaller emitters to provide analysis that captures about 85% of global emissions and 70% of the global population. Crucially, its concern is with the likelihood of nation states' NDCs failing to align with the *Paris Agreement* in terms of policies and actions that are being implemented (CAT, 2024). Chapter 4 in part reviews CAT and CCPI analysis of a selection of EA&P countries, in terms of our climate leadership criteria of credibility and effectiveness, responsible and coherent action, and of actions to address the differing developed, lesser-developed and developing nation capacities.

If a nation state is doing well against its pledged NDC, in terms of emissions reductions that are being achieved from policy implementation, this provides one means of recognising performance and climate leadership. Leadership on mitigation efforts both domestically and regionally could be expected, for example, from the wealthier EA&P countries, Japan, Australia, and New Zealand, which are each OECD and Annex 1 developed countries. OECD countries are typically high-income countries, whilst the UNFCCC Annex 1 includes both developed countries and those transitioning to a developed status that have agreed to take, record and report mitigation action under UNFCCC conventions, protocols and agreements. However, whilst national reporting to the UNFCCC shows that Japan has, for example, reduced its CO_{2e} emissions by 7.98% on 1990 levels, the developed countries of Australia and New Zealand have increased their emissions since then by 24.09% and 18.70%, respectively (Table 1.2). In this book, we scrutinise this performance against our criteria in looking for regional climate leaders and leadership potential.

In Chapter 4, in part, we review the CAT findings that the climate policies, actions, targets and financial efforts of developed countries such as Japan and Australia are 'insufficient', and New Zealand is 'highly insufficient' to meet the goals of the *Paris Agreement* and therefore to avert dangerous climate change (CAT, 2024; Chapter 4). Collectively, these countries account for only 3.41% of global emissions (Table 1.2), but CAT finds that, for varying reasons, their climate policies and commitments need substantial improvements. Clearly, these countries are not showing leadership in a region where lesser-developed and developing country emissions have been burgeoning since 1990, for example, in China by 203%, Indonesia by 107% and Viet Nam by 229% (Table 1.2). Given such exponential increases, and the population levels and rapid industrialisation in these lesser-developed and developing countries, we include a selection of them in our Chapter 4 country analysis of developed, lesser-developed and developing nation climate leadership against our regional climate leadership criteria. Table 1.2 here also shows that the bulk of EA&P emissions are nevertheless generated by China (Chapter 7).

Whilst the combined emissions of Japan, Australia and New Zealand are a fraction of the global total (Table 1.2), we suggest that effective global emission reduction and adaptation efforts need the efforts of developed nations to 'make things happen' (Huxham, 2003, p. 245). Leading actions by Japan, Australia and New Zealand would not only lift their respective national domestic performances but also influence the regional context by assisting lesser-developed and developing countries potentially by building their capacity. To demonstrate climate leadership in this extended sense would be to undertake effective actions to reduce emissions and to adapt to climate change both domestically ***and*** within the region, and to encourage and assist others to do so (Crowley & Nakamura, 2015; 2017). Leadership within a region, or action by regional agreement or consensus, would not substitute for domestic performance, but would complement it and, ideally, would stimulate additional initiatives, thereby contributing to the polycentric, multi-level governance efforts that Ostrom (2010) suggests are required to help realise action of global significance on climate change.

We argue that climate leadership within the EA&P region is a more nuanced concept, then, than simply the leading emission reducing actions of a developed country. We are interested in demonstrated leadership not only in the national context but also within and by the region, and indeed by partners within the region, given that regionally specific mitigation efforts would likely build resilience and capacity and accelerate mitigation. We are also interested in probing beyond climate policy efforts, domestically and within the region, to consider the credibility and effectiveness of such efforts. And we are interested in the neighbourly, regional context of polycentric policy actions, with an emphasis on distributional fairness, and the need for action to be coherent, rather than undermining other actions. Lastly, we are interested in capacity and differentiation as criteria for explaining the differing policy capacities of developed, lesser-developed and developing nations and the capacity challenges facing the latter (Chapter 3). Definitions of climate leadership that are agnostic about regional context and synergies are less helpful for our research; however, we do consider the extent to which they may complement the types of regional leadership that our analysis generates (Chapter 8).

About This Book

In this book, we recognise the challenge of responding to climate change in the EA&P region and argue that regional climate leadership can be seen in this case as a nuanced, polycentric, multi-levelled notion that promotes facilitative actions in a complex neighbourly context. The aim of such leadership would be to lift domestic performance whilst fostering collaborative and regional efforts, predominantly in terms of nation state action, but also potentially in terms of the actions and determinations of regional and sub-regional forums, dialogues and frameworks (Table 2.1). In Chapter 1, we adopt a definition of EA&P that includes developed, lesser-developed and developing countries (Table 1.1) and which is broad and complex in geographical scope. We introduce several key themes: that EA&P is vulnerable in the extreme to climate change; that responding to climate change impacts is daunting in terms of the urgency, scale and capacity that is needed; and that polycentric action at multiple levels, and in multiple arenas and sectors, is required.

In Chapter 2, we detail the complexity of EA&P, the challenge of climate change within the region, the varying climate policy efforts and responses and the complexity of domestic and regional governance as a means of advancing polycentric climate solutions. We review climate change and its impacts in terms of regional, geographical, political and organisational challenges and consider the varying climate policy capacities and vulnerabilities of the countries within EA&P. This leads us, in Chapter 3, to an examination of climate leadership and governance literature to inform and develop our framework for assessing regional climate leadership that is specific to EA&P and focused upon cooperative action. We distil the criteria for analysing regional leadership in terms of the credibility and effectiveness of domestic actions, the coherence and responsibility of action within the region and the capacity and actions of differing types of nations.

Chapter 4 provides a climate policy performance profile of select developed, lesser-developed and developing EA&P countries, and discusses why policy action, policy efficacy and climate leadership by these countries matter in a regional context. It compares policy efforts, domestically and within the region, by employing our criteria, firstly by detailing and comparing our selection of countries within the EA&P region. Our comparative criteria include committing to act; setting domestic targets; adopting effective instruments; not relying on off-setting; and transitioning to renewable energy; and, more broadly, accepting regional responsibility; accepting differentiated responsibility; cooperating on regional climate action; negotiating with major regional emitters; and acknowledging humanitarian need. We conclude by discussing the varying capabilities within EA&P, the importance of regional efforts and the key role that leadership and differentiated responsibility could play in accelerating action.

In Chapters 5 and 6, we extend this discussion by considering the dual 'domestic-regional' nature of the climate policy challenge in EA&P firstly in terms of managing risks and then in terms of implementing carbon pricing domestically and within the region. Both chapters illustrate the complexity of the challenges facing EA&P countries in terms of both managing and reducing climate risks in a manner that aligns with the goals of the *Paris Agreement*. We argue in both chapters that the differing regional forums, dialogues and networks in EA&P have a proven ability to negotiate disaster preparedness, adaptation planning and crisis responses. There are pre-existing regional pathways to complement domestic actions in developing and promoting cooperative climate solutions that can navigate the region's complexities and offer sub-regional solutions. However, lesson learning with a clear focus on the efficacy of climate policy implementation is critical. We also stress that carbon pricing, carbon trading in particular, is a useful, but flawed emissions reduction approach that must be embedded with a context of supportive complementary and regional policy.

Our regional climate leadership approach raises the issue of the leading and lagging actions of China as an emergent superpower and climate leader within the EA&P, albeit one which is still a developing country. China deserves its own chapter for a variety of reasons. It is now the world's largest GHG emitter, responsible for a third of global emissions, and it is now the world's second largest economy, so it has the financial capacity to act on climate change. In Chapter 7, we again employ our leadership criteria to review the climate policy efforts of China, domestically and within EA&P, and we identify a complex, contradictory record whereby China's growth, security and territorial influence all play a significant role in climate policy. In the chapter, we see how China's role in global climate action has evolved from bystander to leader, how its own climate policy and institution building have accelerated and how it is now the largest manufacturer and supplier of renewable energy. But China's role is Janus-faced as a climate leader and a climate villain because of its struggles to move beyond fossil fuels.

In Chapter 8, we revisit the threat of climate change in EA&P and our argument that regional climate leadership is urgently required. We review our country comparisons in terms of identifying regional climate leaders and leadership and consider what our findings add to the climate leadership debates. Based upon our analysis, we refine our definition of regional climate leadership to include domestic actions being complemented by actions taken with neighbourly intent, and regional actions that are supportive of building resilience, reducing vulnerability and leveraging EA&P emissions reduction capacity. We revisit the lessons learnt about regional capacity and policy effectiveness from EA&P cooperation in terms of disaster preparedness, resilience building and various efforts at carbon pricing. We suggest that such lessons are relevant to the development of collaborative action. We conclude that climate leadership that achieves accelerated climate action is possible within EA&P but that it must be reconceived with regional intent.

Conclusions

Climate leadership is typically recognised as the leading actions by developed countries in the multilateral context, often, but not always, within Europe, or led by the EU (Schreurs & Tiberghien, 2007). There is some emphasis on leadership by the United States, and on the emergent leadership of China, that recognises both breakthroughs and failures; and on leading actions that may nevertheless be rife with contradiction. Climate leadership is often seen as 'what works best' or 'which country has the best performance' or negotiating power on the international stage, offering role models for successful action and climate solutions. We argue that the EA&P context calls for a more nuanced understanding of leadership, by introducing regional, indeed sub-regional, leadership, which recognises the building blocks that need to be constructed of domestic, neighbourly, and regional actions. European notions of leadership, cooperation and action do not readily transfer to EA&P, and our book is novel, not only for defining regional climate leadership specific to this context but also for identifying how its fractured regional architecture can respond to climate change in terms of both urgent and long-term solutions, despite its vastness and complexities.

Our contributions include re-framing the dialogue about the urgent emissions reduction that is required to meet the goals of the *Paris Agreement*, and to avert dangerous climate change, in the context of EA&P and the vastly differing country capacities within it. To assess these differing capacities, we re-frame climate leadership as cooperative and facilitative rather than as simply competitive and define dual 'domestic-regional' obligations to act that extend responsibility for climate action into the EA&P neighbourhood and region. We furthermore define regional climate leadership as a synergistic concept, whereby the actions of a developed nation may, for example, facilitate and/or accelerate the actions of lesser-developed and developing countries, thereby lifting the performance of the region. Whilst we see such actions at times as occurring within the context of regional forums, dialogues and frameworks or agreements, we do not attempt to describe an ideal regional climate policy architecture or body akin to the European Union. We remain focused on climate policy and urgent solutions.

Notes

1 The Climate Change Performance Index performance ranking criteria include GHG emissions (40%), renewable energy (20%), energy use (20%) and climate policy (20%).

2 The Republic of China (ROC) was based on mainland China from 1912 to 1949 before its government fled to Taiwan. It is established as an autonomous, democratic region of China, rather than as a country.

2 The Climate Challenge in EA&P

Introduction

Climate change is challenging within the East Asia and the Pacific region (EA&P) on many levels. Fast, effective action to reduce emissions and to adapt to climate impacts is urgently required. In Chapter 1, we argued that regional climate leadership, which promotes domestic, neighbourly and regional action, could accelerate mitigation and adaptation efforts across EA&P. However, we also identified the paradox that the developed nations in EA&P most able to assist their lesser-developed and developing nation neighbours are performing poorly in terms of reducing their own emissions. In Chapter 3, we therefore offer a set of normative criteria for offering an assessment of the aspects of leadership that the climate policy, leadership and governance literature identify as crucial to improved performance and to assuming a climate leadership demeanour. In this chapter, we argue that the context for assuming and inspiring climate leadership is an extremely complex one in EA&P, where building blocks are required of transferable, effective, polycentric action. Region, leadership and polycentrism are therefore each significant aspects of the challenge of climate policy development, governance and accelerated action within EA&P.

In this chapter, we detail the complexities of EA&P, with very clear differences, for example, between its regions, sub-regions and relationships. The vast EA&P region does not have unified, structured, rules-based, regional governance comparable to the European Union (EU), with the ability to act as a centralising climate policy developer and driver towards enhanced monitoring, implementation and outcomes evaluation. Instead, regionalism is a more dynamic, more contingent project in EA&P. Indeed, Dent (2008, p. 3) argues that it is the various integrative economic, political and sociocultural processes that define regionalism in EA&P rather than any institutionalised structures of central governance. Climate leadership in this fractured context is, as we argue in Chapter 1, a nuanced concept. In this chapter, we review the complexity of EA&P, the challenge of climate change in the region, the varying policy efforts and domestic responses and the regional frameworks, agreements and dialogues that facilitate climate policy solutions.

DOI: 10.4324/9781003170389-2

The Complexity of EA&P

Conceptions of EA&P vary and reflect the differing interests of nation states, economic interests and actors, supranational organisations and governance, and trading and security arrangements (Table 1.1). The United States Bureau of East Asian and Pacific Affairs (BEAPA) defines a broad, sweeping region eastwards from China, across the East Asian seas and the Pacific and stopping at its own borders, but the United States nevertheless lobbies for recognition as an EA&P nation where it suits its interests. The World Bank, headquartered in Washington, DC, describes an EA&P region that resembles the BEAPA's definition, in its partnering within the region on promoting and financing climate change action. It describes a region that accounts for one third of global greenhouse gas (GHG) emissions and 60% of the world's coal consumption, with 13 of the 30 countries that are most vulnerable to environmentally destructive and poverty inducing climate change impacts (World Bank, 2023a). The inclusion of Pacific Island countries, which are strategically significant for the United States, and significant in terms of our interest in regional climate leadership as countries within the regional neighbourhoods of Japan, Australia and New Zealand, is notable in both the BEAPA and World Bank definitions.

Definitions of the Asia Pacific nevertheless vary considerably, again according to the politics, trading arrangements, agendas and involvements at stake. Whilst countries within and bordering the Pacific Ocean are typically included, they are not always. Depending on the circumstances, the Americas may be included or excluded, as may Eastern, Western and Southern Asian, and Indo-Pacific, countries. Developed countries may be included or excluded, as may dependent territories, and the inland countries of East Asia (World Population Review, 2023). The United Nations Economic and Social Commission for Asia and the Pacific (UNESCAP, 2023b) offers the broadest definition of the Asia Pacific. Well beyond our focus on EA&P, this definition includes a vast range of countries like Russia, Turkey, Iran, Afghanistan, Uzbekistan, Kazakhstan, Nepal, Sri Lanka, North Korea and Southern and Northern Asian countries. UNESCAP partners with such countries at the national and regional levels in supporting knowledge generation, technical development and capacity building of direct relevance to promoting effective climate action that is significant for facilitating action in EA&P. More typical are the East Asia and Asia-Pacific definitions, and the recognition that the former 'looks set to be dominated by a resurgent China' and the latter by the USA which is unambiguously already a global power (Beeson, 2009, p. 1).

In terms of East Asia, Dent (2008, pp. 4–12) observes that there are regions within the region, but that East Asia 'has also become a more coherent entity as a result of deepening regionalism and regionalisation'. While there is no formal regional structure offering leadership such as the EU, regional, economic, and organisational arrangements, dialogues, organisations, frameworks and sub-regional structures remain at least potentially significant. Dent also observes that there has been an inevitable rise of dominant regional powers, like Japan and China, rather than a regional representative body, and more recently an increase in security and trading tensions between these two countries. Rather than being defined by Dent's conception of coherence – in terms of regionalism, association and integration – East Asia's 'regional community building' remains weak except for the Association of Southeast Asian Nations (ASEAN). Dent describes its economic development as asymmetric. Its political regimes, traditions and characteristics are diverse, and conflict and nationalism persist (Dent, 2008, p. 10).

East Asia is also drawn into a broad range of regional organisations and frameworks, many of which include the Pacific (Table 2.1). Table 1.2 identifies the differing financial capacities and emissions impact of EA&P countries and in Chapter 4 we discuss the differing capacity of EA&P countries to act on climate change. The countries of East Asia are vastly wealthier and more populous than those of the Pacific and are predominantly responsible for the region's emissions, rather than the Pacific countries, with the exceptions of Australia and New Zealand, both of which rate as significant climate policy laggards (Chapter 1).

The Pacific Island countries differ significantly from those of East Asia not least in terms of the lack of deepened regionalism and regionalisation to which Dent refers but also in terms of population, GDP, per capita emissions, their contribution to global emissions and the rate of emissions growth, which remains negligible in these countries (Table 1.2). Pacific countries are considered regional neighbours of Australia, New Zealand, the United States and to some extent Japan, and our definition of regional climate leadership includes an expectation of support from these developed nations to assist mitigation efforts not only by reducing their own emissions but also by helping these smaller countries to develop resilience and to adapt to the impact of climate change (see Chapters 3 and 5). This expectation has become more complex as China has sought to enhance its regional influence by expanding its aid to Asia Pacific countries, including aid for building climate resilience (Zhang, 2022; Chapter 7).

The Pacific Island countries are included in the UN's definition of small island developing states (SIDS), which it defines as follows:

> … low-lying coastal countries that share similar sustainable development challenges, including small population, limited resources, remoteness, susceptibility to natural disasters, vulnerability to external shocks, and excessive dependence on international trade. Their growth and development is often further stymied by high transportation and communication costs, disproportionately expensive public administration and infrastructure due to their small size, and little to no opportunity to create economies of scale.
>
> (UNESA, 2007)

Given their vulnerability, the SIDS of the Pacific, together with those of other regions,[1] have demonstrated leadership in climate change action by forming the Alliance of Small Island States (AOSIS) and by playing a key

Table 2.1 EA&P, regional climate policy pathways, organisations, strategies and statements

Organisation	*Countries*	*Climate policy pathways and initiatives [examples]*	*Policy areas of interest*
ASEAN countries	Brunei, Cambodia, Indonesia, Myanmar, Lao PDR, Malaysia, Philippines, Singapore, Thailand, Viet Nam	ASEAN Leaders' Statement on Joint Response to Climate Change 2020; ASEAN Vision 2025; ASEAN State of Climate Change 2021; ASEAN Action Plan on Joint Response to Climate Change (AAP-JRCC); ASEAN Working Group on Climate Change; ASEAN Climate Resilience Network	Includes: equity; stabilising emissions; developed countries leading in mitigation and in assisting developing countries; and in building resilience in ASEAN countries; addressing international issues of common concern such as climate change, environment and sustainable development
ASEAN Plus Three (APT)	ASEAN *plus* China, Singapore and South Korea	See the details in the ASEAN Plus Three Cooperation Work Plan 2023–2027. Support for the establishment of an ASEAN Centre of Excellence for Clean Coal Technology. ASEAN Strategy for Carbon Neutrality	Includes: mitigation; adaptation; the transfer and diffusion of new technology; financial resilience against climate shocks and disasters; climate-friendly bioenergy; climate smart agriculture; clean coal; energy transition; and building climate change resilience
ASEAN Plus Six (ASEAN+6)	ASEAN *plus* China, India, Japan, South Korea, Australia and New Zealand	Bilateral statements, e.g. with China 2021; India 2022; Japan 2022 [ASEAN Japan Climate Change Action Agenda]; ROK (2016–20); Australia 2022; NZ (2021–25). With support for ASEAN regional research and policy institutes	Includes: enhancing and promoting cooperation on disaster prevention, mitigation, capacity building and relief; providing climate-based finance; and support for an ASEAN Centre for Climate Change

(Continued)

Table 2.1 (Continued)

Organisation	*Countries*	*Climate policy pathways and initiatives [examples]*	*Policy areas of interest*
ASEAN Regional Forum (ARF)	ASEAN *plus* ASEAN Dialogue Partners Australia, Canada, China, the European Union, India, Japan, New Zealand, Korea, Russia and the United States; Bangladesh, South Korea, Mongolia, Pakistan, Sri Lanka, Papua New Guinea and Timor-Leste	The ARF facilitates dialogue (multilateral and bilateral) and consultations aimed at securing the maintenance of peace, security and cooperation in the Asia Pacific ARF *Inter-Sessional Meetings on Disaster Relief (ISM on DR)* aim to strengthen cooperation on disaster management and are held annually (first convened in 1997)	Includes: strengthening cooperation on disaster management, accounting for the region's vulnerability to natural disasters, climate change and rising sea levels; enhancing capabilities and capacities of humanitarian assistance and disaster relief efforts; and facilitating expeditious and coordinated deployments by ASEAN Member States' militaries in times of disaster
East Asia Summit (EAS)	ASEAN *plus* Australia, China, India, Japan, New Zealand, South Korea, Russia and the United States	2019 *EAS Leaders' Statement on Partnership for Sustainability* including promoting disaster risk reduction; 2021 *EAS Leaders' Statement on sustainable recovery*	Includes: growth strategies to help finance climate actions; noting efforts to meet *Paris Agt** targets; promoting cooperation; and building climate resilience
Asia-Pacific Economic Cooperation (APEC)	Australia; Brunei Darussalam; Canada; Chile; People's Republic of China; Hong Kong, China; Indonesia; Japan; South Korea; Malaysia; Mexico; New Zealand; Papua New Guinea; Peru; the Philippines; the Russian Federation; Singapore; Chinese Taipei; Thailand; the USA; Viet Nam	2020 *APEC Putrajaya Vision 2040* (including actions for strong, balanced, secure, sustainable and inclusive growth); 2015 Cebu Declaration *Towards an Energy Resilient APEC Community*	Includes: ensuring resilience; skill development; and knowledge production; as well as supporting efforts to address climate change, extreme weather and natural disasters; and determining the appropriate fuel and power generation technology mix that would support the twin goals of economic prosperity and environmental sustainability

(Continued)

Table 2.1 (Continued)

Organisation	*Countries*	*Climate policy pathways and initiatives [examples]*	*Policy areas of interest*
Pacific Islands Forum	Australia, Cook Islands, Federated States of Micronesia, Fiji, Kiribati, Nauru, New Zealand, Niue, Palau, Papua New Guinea, Republic of the Marshall Islands, Samoa, Solomon Islands, Tonga, Tuvalu and Vanuatu	*2050 Strategy for the Blue Pacific Continent; Pacific Resilience Partnership* – established by PIF Leaders in 2017 to implement the *Framework for Resilient Development in the Pacific: An Integrated Approach to Address Climate Change and Disaster Risk Management 2017–2030 (FRDP); The Pacific Sustainable Development Report 2020*	Includes: urgent, immediate action to address climate change; reducing and preventing climate change impacts; enhancing carbon sequestration; scaling up effective and sustainable climate finance; ensuring resilience and future-readiness; addressing inefficient energy use; strengthening regional cooperation; protecting rights and addressing gendered climate impacts; investing in research
SPC Pacific Community	Cook Islands, Fiji, Kiribati, the Northern Marianna Islands, New Caledonia, Palau, Papua New Guinea, the Pitcairn Islands, French Polynesia, Tokelau, Tonga, Tuvalu, Niue, the Marshall Islands, the Federated States of Micronesia, Wallis and Futuna, Nauru, the Solomon Islands, American Samoa, Vanuatu and Samoa; and founding countries Australia, France, New Zealand, the USA and the United Kingdom	*Taking Action on Climate Change to Shape a Resilient Pacific 2023* The SPC Pacific Community is the principal scientific and technical organisation in the Pacific region. It is an international development organisation owned/governed by member countries and territories. It works in cross sectoral fashion in pursuit of sustainable development to the benefit of Pacific people	Includes: actions to achieve the objectives of the *Paris Agt*; and actions supporting the building of resilient sustainable development; addressing (1) *climate action* – mitigation; loss, damage, climate justice; adaptation and resilience; and (2) *enabling action* – monitoring, reporting; policy, advocacy, leadership; science, information, knowledge; and climate finance

(Continued)

Table 2.1 (Continued)

Organisation	*Countries*	*Climate policy pathways and initiatives [examples]*	*Policy areas of interest*
Pacific Economic Cooperation Council (PECC)	Australia; Brunei Darussalam; Canada; Chile; China; Colombia; Ecuador; Hong Kong; Indonesia; Japan; South Korea; Malaysia; Mexico; Mongolia; New Zealand; Peru; the Philippines; Singapore; Taiwan; Thailand; the United States; Viet Nam; the Pacific Island Forum; Associate member France (Pacific Territories), and Institutional members, the Pacific Trade and Development Conference (PAFTAD) and the Pacific Basin Economic Council (PBEC)	2022 *State of the Region* report; 2020 *APEC Putrajaya Vision 2040* (including actions for strong, balanced, secure, sustainable and inclusive growth)	Includes: promoting resilience in the context of sustainable growth; knowledge and skills development, economic policies, cooperation & growth which support global efforts to comprehensively address all environmental challenges, including climate change, extreme weather & natural disasters, for a sustainable planet. Climate change as a risk to growth so cooperative, regional efforts required to reduce the risk (noting that it is difficult to separate out cooperation from competition in other areas)
UN Small Island Developing States – Pacific	Fiji, Maldives, Marshall Islands, Federated States of Micronesia, Nauru, Palau, Papua New Guinea, Samoa, Singapore, Tonga, Vanuatu, Kiribati, Solomon Islands and Timor-Leste	The UN Pacific Climate Change Migration and Human Security Program (PCCMHS) – delivered through a partnership between PIF and the UN Agencies of IOM, ILO, OHCHR and UNESCAP	This program seeks to protect/empower communities affected by climate change and disasters in the Pacific region, focusing specifically on climate change and disaster-related migration, displacement and planned relocation

Sources: APEC (2015, 2023); ASEAN (2010, 2021, 2022a, 2022b, 2023a; 2023b, 2023c); DFAT (2022); PECC (2021a, 2021b); PIF (2022); PRP (2023); SPC (2022).
Note: *Paris Agt – Paris Agreement.

role in international negotiations on global action on climate change throughout the *Kyoto Protocol* and *Paris Agreement* phases over recent decades (UNFCCC, 2005). AOSIS's lobbying and campaigning, including their more recent '1.5C to Stay Alive' efforts, have been crucial in setting the *Paris Agreement's* goal of reducing emissions to achieve – as nearly as possible – no more than 1.5 °C warming by 2050 (Thomas et al., 2020). These efforts by Pacific Island SIDS are examples not of unilateral leadership by a nation state in terms of our own leadership criteria, but of collective regional climate leadership in Dent's (2008, p. 22) terms of leading regional community building, resolving collective action problems and championing and representing the interests of the region in the wider global community.

Regional Climate Governance and Frameworks

Collective regional action on climate change requires supportive regional governance mechanisms and frameworks, but also unanimity of purpose and a problem-solving mindset that is unhampered by national interests, or, conversely, that is driven by a shared experience of climate change impacts. The EA&P governance and policy context is vast and fractured, and therefore contingent, in contrast to the unifying context afforded by the EU; however, there are a plethora of formal and less formal pathways for promoting climate action, mitigation and adaptation efforts. These vary considerably in terms of membership, style, scope, longevity and likely effectiveness. The most significant governance groupings best suited to driving regional, rather than country-based, action on climate change in EA&P are the ASEAN and the Pacific Island Forum (PIF). Beyond these organisations are the predominantly East Asia based extended ASEAN groupings, ASEAN Plus Three (APT), ASEAN Plus Six (ASEAN+6), the East Asia Summit (EAS) and the ASEAN Regional Forum (ARF). The Pacific countries of New Zealand and Australia feature as members of several of these. Additional bodies with overlapping East Asia and Pacific membership include the economic and trade focused Asia-Pacific Economic Cooperation (APEC) and its forebear, the Pacific Economic Cooperation Council (PECC). In the Pacific, in addition to the PIF, is the broad Pacific Community grouping, which includes previous colonising countries, and the United Nations facilitated SIDS – Pacific grouping which includes the lesser-developed Pacific countries (Table 2.1).

East Asia Regional Climate Policy Focus

ASEAN is a relatively narrow regional grouping, but it has evolved into an inter-governmental organisation of developing nations with a charter registered with the United Nations that details its legal and institutional framework, and its norms, rules, values, targets, and accountability and compliance procedures. It has developed clear climate policy implementation pathways. The *ASEAN Vision 2025* includes a high-level commitment to addressing climate change (ASEAN, 2015) that builds on earlier progress. In 2010, ASEAN Leaders released the *Statement on Joint Response to Climate Change* and established an *Action Plan on*

Joint Response to Climate Change (AAP-JRCC) and a Working Group on Climate Change. This was followed by the launch of the first ASEAN State of Climate Change Report. This 'provides an overview of the region's status on climate capacity, outlines how actions can be further improved, as well as identifies opportunities for cooperation and collaboration to support ASEAN's efforts towards achieving the 2050 net zero transition targets' (ASEAN, 2020). This report notes ASEAN's commitment to net-zero emissions and to emissions peaking soon after 2030, but also to the enhanced capacity building that is required to decarbonise effectively and that must now be the focus of ASEAN and its member states.

Further climate policy and capacity building leverage is possible through APT, which includes China, Singapore and the Democratic Republic of Korea (South Korea), and ASEAN Plus Six (ASEAN+6), which includes China, India, Japan, South Korea, Australia and New Zealand. The EAS, which adds the United States and Russia to ASEAN+6, is a significant leader-only grouping, representing 54% of the global population and generating 62% of global GDP, with a focus on strategic dialogue and regional cooperation (DFAT, 2022). ASEAN countries are concerned that regional climate policy focuses on equitable efforts between developed, lesser-developed and developing nations and that, in stabilising emissions, developed countries take responsibility for leading action, and for assisting in building resilience in ASEAN countries (ASEAN, 2010). Such responsibility features in our approach for assessing regional climate leadership (Table 3.3) and could be reflected within bilateral statements and agreements between ASEAN and developed nations (Table 2.1). ASEAN's extended partnership forums offer the means of embedding equitable regional decarbonisation, capacity and resilience building goals, policies and practices into regional agreements, statements and processes.

Pacific Regional Climate Policy Focus

The Pacific region has a cohesive institutional mechanism in the PIF, but PIF operates in a style that is less formal, and less integrated than ASEAN, and thus less likely to foster the ASEAN trend towards deepening regionalism and regionalisation. ASEAN itself is in no way as cohesive and rules based as the EU, but it has agreed viable climate policy implementation pathways, which nevertheless remain reliant upon the capacity, inclinations and leadership of individual member countries (see Seah & Martinus, 2021). Gammon (2022) argues that the 'building blocks' of Pacific regionalism are not as strong overall as in South-East Asia because of the remaining development challenges for PIF member countries and that they must therefore engage, ASEAN style, with outside countries to build their capacity (Powles & Wallis, 2022). As well as building capacity, the building of unity amongst PIF member countries and partners is key to tackling threats, including the impacts of climate change in the region, and to progressing decarbonisation practices. Having debated its lack of effective regional unity (PIF, 2019) as a threat to the interests, national security and in some cases the long-term viability of Forum Member countries, PIF recently endorsed the development of a climate change oriented *2050 Strategy for the Blue Pacific Continent* (PIF, 2022).

The context for this strategy is that PIF members and partners are stewards of an increasingly environmentally threatened and geopolitically contested Pacific Ocean and that deepening Pacific regionalism with a robust forward-looking approach is the required antidote. The individual nation states and the broader groupings in the region, for example, the SPC Pacific Community, the PECC, and the Pacific chapter of the United Nations UN SIDS (Table 2.1), now have a clear point of orientation for aligning and integrating their climate policy efforts with the PIF's. The strategy declares values and aspirations for regionalism, and in its task orientation, and has been designed for policy learning, in terms of building upon existing decisions and frameworks, and with effective implementation. 'Climate change and disasters' is a key thematic area. For each thematic area, there are strategy 'pathways' – in terms of governance, inclusion and equity, education, research and technology, resilience and well-being, and partnership and cooperation. The next steps will determine an implementation plan for each thematic area, collective member country actions, delivery time frame, implementation arrangements, partners and required resources with the intent to align national, regional and global efforts to address climate policy goals (PIF, 2022).

Regional Climate Policy Pathways

Table 2.1 details various regional policy pathways, including ASEAN groupings plus the ARF and the EAS, but also the climate policy focus of the economic and trade focused APEC, and of regional bodies in the Pacific. There are significant opportunities within these policy pathways, initiatives and fora for a member state to demonstrate leadership by negotiating an escalation of regional climate policy ambition, an acceleration of emissions reduction actions and, potentially, a regional emissions reduction target.

Climate Policy Challenges

The climate policy challenges facing EA&P are indeed complex, and differentiated, with East Asian countries more populous, wealthy and emission intensive than the countries of the Pacific, and with climate impacts already eroding the shorelines of Pacific Island countries (Cassella, 2019; Chapter 4). However, the climate policy adoption and institution building in ASEAN and PIF over the last decade show the potential for collective problem-solving on a regional basis in EA&P and have seen the development of regional climate working groups, statements and work plans. In turn, the embedding of mitigation, adaptation and capacity building processes into regional governance creates a context, in terms of broad goals and aspirations, to guide the efforts of member states. Regional vision, integrated strategies, implementation pathways and the unity of purpose achieved by the ASEAN and PIF processes have the potential to foster accelerated climate action. But will this potential be realised, and will that be fast enough to accelerate the mitigation needed to avert, and to build resilience against, the dangerous impacts of climate change within the region? The national interests served by reliance upon fossil fuels remain a major obstacle, with carbon-neutral pledges and rapid advances in the uptake of renewable energy undercut in China, for example, by its recent ramping up of coal production (Hawkins, 2023; Chapter 7).

The Coal Challenge

The EA&P remains endemically reliant upon coal (World Bank, 2023a). In 2023, the EA&P region accounted 'for more than half of global energy consumption, with 85% of that regional consumption sourced from fossil fuels' (IRENA 2023). By the end of 2022, the region had 42.5% of proven coal reserves globally (China 13.3%), 77.3% of total global coal production (China 50.8%) and 79.7% of the global total consumption (China 53.8%). In terms of the percentage of global coal reserves in the region beyond China, Australia had 14%, Indonesia 3.2% and, less notably, New Zealand had 0.7%, Thailand 0.1%, Viet Nam 0.3%, whilst the rest of Asia and the Pacific only 0.2%. In terms of percentage of global consumption in the region beyond China, Australia had 7.4%, Indonesia 9% and, again less notably, Thailand 0.1, Viet Nam 0.7% and the rest of Asia and the Pacific only 0.7%. Totals for coal production in the region beyond China were generally well less than 2% (Energy Institute, 2023). Several EA&P countries also rank within the top 20 countries in terms of cumulative (billion tons of CO_2 from fossil fuels, cement, land use and forestry) emissions from 1850 to 2021, i.e. China (2nd); Indonesia (5th); Japan (9th); Australia (13th); and Thailand (18th) (Carbon Brief, 2021).

As well as being overly reliant upon fossil fuels, the Asia Pacific is expected to see growth to 2050 in its emissions at twice the global average, driven by rapid industrial growth and the dramatic population growth that is expected in the region (WRI, 2023). Drastic action to shift from coal and to accelerate a region wide transition to renewable energy sources will be needed to meet its contributions to global emission reduction targets by 2050 (Romsom & McPhail, 2020). The challenge will be for Asia-Pacific countries to align their growth expectations with their transition efforts, and not to hamper renewable energy growth in the region by failing to phase out fossil fuels. Coal consumption trends need to reverse, most notably in China and India, which are the region's biggest emitters, and coal production must be reined in by the region's top coal producers, China, Indonesia, India, and Australia (Figure 2.1). However, the International Monetary Fund reports that coal generation is instead expanding, with the Asia-Pacific region accounting for virtually all the global pipeline for newly planned coal plants, and with the region's coal-fired power generation expected to continue at a very high level, phasing out only around 2060 (IMF, 2021a).

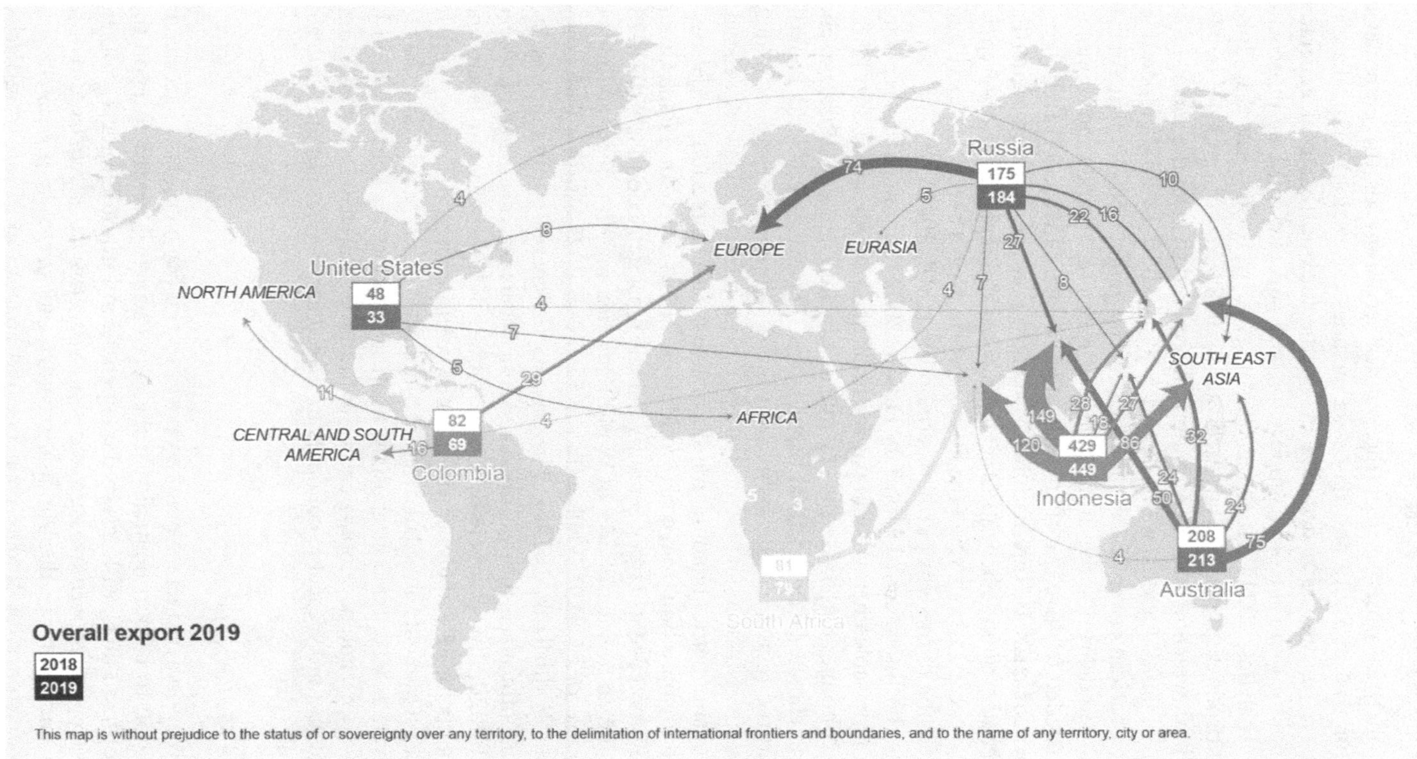

Figure 2.1 Main trade flows in the thermal coal market, 2019 (Mt) – showing Indonesia and Australia as the world's top coal exporting nations in 2019.

Source: International Energy Agency (2020a) (available under Creative Commons licence).

Finance and Partnership Challenges

Climate Analytics observes that the political economy of the entrenched support for coal in South and East Asia includes the region's large share of coal reserves, its historical dependency on the income that coal generates and the political and financial support for, and subsidies of, coal in general (Climate Analytics, 2021). However, Climate Analytics also warns that global fossil fuel production must be cut each year by 6–8% to reduce its use by 40–45% by 2030, which would be the minimum required to meet the *Paris Agreement* goal of limiting global warming to 1.5 °C by 2050. By 2030, renewables should be providing over 70% of global electricity generation (Climate Analytics, 2023) and delivering the sustainable jobs, green technology and economic stimulus needed to drive EA&P growth and resilience (IMF, 2021a). In terms of cost-benefit analysis, International Monetary Fund researchers argue that the social and economic benefits of scrapping coal and replacing it with renewable energy would far outweigh the financial costs, with a net global social gain estimated to be in the tens of trillions of dollars (IMF, 2022).

Despite the International Monetary Fund's (2022) argument that replacing coal with renewable energy could pay for itself, the short-term costs to EA&P country economies emphasise the key role of climate finance and bilateral and polycentric climate-based financial partnerships. For example, World Bank financing in the region already assists with the development of carbon trading and emissions reduction and includes support with resilience financing, climate diagnostics and metrics building, transformative investments and financial assistance with the transition from coal (World Bank, 2021). The World Bank Group has already exceeded its target, set in its Climate Change Action Plan 2021–25, of deploying an average of 35% of World Bank funding on climate action and spending a record US$31.7 billion in the fiscal year 2022 assisting with country action (World Bank, 2022b). The capacity for energy transitioning is also robust currently, given that wealth in the Asia Pacific has quadrupled between 2000 and 2020, driven largely by China but also by Australia and Japan, and now accounts for 42% of global wealth, and therefore offers the opportunity to recalibrate investment towards a cleaner economy (Woetzel & Sengupta, 2022).

However, as Chapter 5 establishes, financial support for climate action in lesser-developed and developing countries from the United States, the EU and developed and more capable countries in EA&P is both fraught and critically insufficient.

Transition Challenges for EA&P

The International Monetary Fund proposes several key measures to accelerate the transition to carbon neutrality in the Asia Pacific: (i) *carbon taxes* on the burning of fossil fuels, with targeted compensation for householders, workers and firms; (ii) *complementary regulatory and incentive based policies* to reduce pollution and promote decarbonisation; (iii) and *steepened investment and financial assistance* for climate resilience and adaptation in low-lying Pacific countries (Dabla-Norris et al., 2021). The crucial prior measure is, however, acceptance by Asia-Pacific countries that coal is no longer a viable financial or environmental investment because the cost competitiveness of renewables versus coal is rapidly accelerating (Carbon Tracker, 2021). Carbon Tracker observes that 27% of the global coal fleet is economically unviable and can maintain operations only because of policy support and market distortions and that this figure will reach 66% by 2040. Crucially, it is the actions of China, India, Indonesia, Viet Nam and Japan that could push global carbon neutrality beyond reach since these countries are responsible for 80% of the world's new coal plants and 75% of existing coal capacity. 'In these five countries, 92% of planned coal units will be uneconomic, even under business as usual, and up to $150 billion could be wasted' (Carbon Tracker, 2021).

The UNESCAP has resolved to end the region's dependency on coal by 2050 and to radically phase down oil and gas (UNESCAP, 2023c; see also Table 2.2). It will seek financing deals, such as Just Energy Transition Partnerships (JETPs), to

Table 2.2 The coal challenge in East Asia and the Pacific

Country	*Coal* Reserves/ global ranking*	*Coal** Production/ global ranking*	*Coal** Consumption/ global ranking*	*Coal** Imports/ global ranking*	*Coal** Exports/ global ranking*
China	157847.49/1	4548131/1	4675320.5/1	356794.19/1	2839.91/14
Australia	165596.72/3	514937.16/5	109921.25/11	123.98/81	403224.47/2
Indonesia	38436.45/7	676807.63/3	165481.05/8	14812.38/16	478028.69/1
New Zealand	8350/13	3161/36	3463.17/50	2033.98/43	1371.68/19
Viet Nam	3703.76/19	52740.2/14	101230.62/12	44010.64/5	1070.38/22
Thailand	1171.76/33	15676.58/24	37703.41/18	26531.66/11	167.5/35
North Korea	661.39/36	23391.02/20	8134.83/37	0/169	0/156
Lao PDR	554.46/38	15781.32/23	15409.16/26	30.95/98	403.11/30
Philippines	397.93/44	15849.01/22	28090.17/23	34435.06/10	10287.86/12
Japan	385.81/45	743.74/50	201874.58/5	201137.09/3	6.26/52
South Korea	359.35/48	—	—	—	—
Malaysia	249.12/52	3438.23/35	36224.59/19	38673.22/9	157.91/37
Myanmar	6.61/70	2724.08/38	2532.21/57	16.07/103	1.28/60
Taiwan	1.1/79	0/180	71852.98/14	74296.78/4	1.17/61
Brunei	0/94	0/80	0/134	0/132	0/87
Cambodia	0/97	0/83	2932.98/53	2854.14/36	0/90
Fiji	0/115	0/105	0/148	0/146	0/111
Kiribati	0/137	0/129	0/159	0/157	0/132
Papua NG	0/159	—	0/170	0/170	0/159
Samoa	0/167	0/164	0/177	0/177	0/168
Singapore	0/173	0/170	823.73/76	824.08/60	0.36/65
Solomon Islands	0/174	—	0/181	0/181	0/174
Tonga	0/182	0/182	0/185	0/185	0/183
Micronesia	—	0/145	0/167	0/165	0/145
Tuvalu	—	0/186	0/187	0/187	0/187
Vanuatu	0/189	0/190	0/190	0/190	0/190

Source: https://www.theglobaleconomy.com/

Note: *Million short tons; **Thousand short tons; – no data.

Table 2.3 Coal phase out and energy transition pathways for Asia and the Pacific – Climate Analytics' policy recommendations

1	National governments: Adopt best practice policies – phase out fossil fuel subsidies, adopt carbon pricing, develop renewable energy support, encourage and push shift in investment through green recovery plans.
2	National governments: Move to transformational policies, targets and long-term planning – ratcheting up NDC targets and adopting Long-Term Strategies in line with the *Paris Agreement.*
3	Clear pathways to enable anticipation of change and to avoid more stranded assets – moratorium for new coal, targets and phase out plans for coal before 2040.
4	Financial support and capacity building need to be informed by *Paris Agreement* benchmarks to shift investments at scale in the near term.
5	Regional and international cooperation: Alignment with *Paris Agreement* goals, engaging stakeholders, private sector and civil society.
6	Trans-boundary grid transmission and integration to be achieved to underpin clean energy transition.
7	Financial institutions: Cooperation on sustainable finance, clear policies and transparency.
8	Private sector engagement and multi-party efforts.

Source: Climate Analytics (2021)

ramp up renewable energy and energy efficiency and help phase out coal power plants (Table 2.3), even before the end of their useful life (UNESCAP, 2023c, p. 18). ASEAN agrees that, even with carbon capture and storage (CCS), coal must be phased out in the region by 2050, and that ASEAN Member States (AMS) need to immediately prepare strategies or plans to begin phasing out coal power plants and to switch to renewable/low carbon energy (ASEAN, 2020). By contrast, the PIF's 2050 *Strategy for the Blue Pacific Continent* makes no mention of coal, noting that its region generates just over 1% of global emissions, but that it is at the frontline of the adverse impacts of climate change and needs timely access to scaled-up, effective, finance (PIF, 2022). At the 2023 Pacific Ministerial Dialogue on Pathways for the Global Just Transition from Fossil Fuels, Ministers from Vanuatu, Tuvalu, Tonga, Fiji, Niue and the Solomon Islands called for an equitable transition as soon as possible by way of a global Fossil Fuel Non-Proliferation Treaty. Advice on the legal means of reaffirming, strengthening and codifying such a global fossil fuel phase out will be sought from the International Court of Justice (Boudreau, 2023; FFNPTI, 2024a).

Conclusions

We have seen that the context for climate leadership is a complex one in EA&P. In terms of climate change, we have identified differing contexts between the wealthier, more populous, more emissions intensive and faster growing, faster developing nations of East Asia and those of the relatively poorer, smaller, more vulnerable Pacific Island nations. However, vision statements and strategic planning in both ASEAN and PIF contexts are unequivocal about the need to mitigate emissions in part by transitioning towards a fossil free, renewable energy future in line with *Paris Agreement* aspirations (Table 2.3). Although regionalism is a more advanced practice in East Asia than in the Pacific, we identify a broad range of regional climate governance visions, sessions, strategies, tools, statements, frameworks, action plans and programmes across the entire region in general that support action. These have the capacity to support adaptation action, by resilience building, with an urgent focus on Pacific Nations, and specific mitigation action, including carbon trading, with a focus on fledgling efforts in East Asia (Chapters 5 and 6). Both sets of action hold promise of regional leadership in Dent's (2008) terms of coordination at the regional level.

However, we have also identified problematic nation state circumstances and interests that require a novel style of regional climate leadership, namely leadership in terms of leading domestic actions matched by leading actions within, and for the benefit of, the region. We have seen that AMS have agreed to phase out coal by 2050, and to begin their transition to renewable energy immediately, and that the Pacific Island community is pursuing a fossil fuel non-proliferation treaty. But with coal use burgeoning in the region (Carbon Tracker, 2021), will this be enough to prevent dangerous climate change? We have argued that leadership is urgently required and that effective nation state climate action is likely to inspire regional lesson learning and policy transfer within states and regional formal and informal networks, potentially building resilience and emission reducing measures and regimes. In Chapter 3, we explore the normative dimensions of climate leadership to establish, in practical terms, the dimensions of, and indicators by which to measure, a regional climate leadership that comprises both domestic and regional nation state actions. In Chapter 4, we consider whether the climate policy performance of select nation states within EA&P to act on climate change is in any way that of a regional climate leader.

Note

1 The SIDS comprise 38 United Nations member states and 20 non-member/associate states in the Caribbean, Pacific, Atlantic, Indian Ocean, Mediterranean and South China Sea.

3 Defining Climate Leadership in EA&P

Introduction

In international global climate change efforts to date, in an application of the agreed principles of common but differentiated responsibilities, developed countries have been expected to lead on emissions reductions and to assist developing nations with their efforts. This has been a failed effort that was critically reviewed in the lead up to the 2015 UN Climate Change Conference (COP21) in Paris in December 2015. Developing nations are now expected to join the effort to avert dangerous climate change; however, we argue that developed nations must consider what actions they can take to assist developing nations within their immediate region. This is in keeping with the *Paris Agreement's* Article 4.4 proposition that 'developed country Parties shall continue taking the lead' in reducing emissions (Wurzel et al., 2021, p. 4). Developed-developing nation efforts within regional or neighbourly contexts could furthermore provide stepping-stones, we argue, to more effective global climate action. The East Asia and Pacific region (EA&P) includes developed, lesser-developed and developing countries and it models, at the sub-national level, the broader global problem of developed-developing country responsibility for, and engagement on, climate change. This is the context in which we develop criteria for regional climate leadership.

The European-based climate leadership theory does distinguish between actions within and beyond the domestic context. However, it is not readily applicable within the more complex, more institutionally fractured EA&P. Previously, we have argued that there is a need to examine strong governance for facilitating more cooperative climate action between developed, lesser-developed and developing countries within the EAP in the absence of a unifying European Union (EU) context (Crowley & Nakamura, 2015, 2017). In this chapter, we distil criteria for analysing regional climate leadership, which we define in terms of the credibility and effectiveness of domestic actions, and the assumption of responsibility, that is coherent in terms of consistent policies, within a regional context, and that is also cognisant of differing country capacities and recognising capability and differentiation principles. Our approach is policy oriented, identifying and analysing contrasting mitigation,

DOI: 10.4324/9781003170389-3

adaptation and engagement efforts in order to establish the potential for, if not the achievement of, regional climate leadership in EA&P. We are not interested in theorising climate leadership notions and characteristics per se, but rather in identifying and developing normative criteria that are useful for reflecting upon regional climate leadership in Chapter 4.

The Climate Leadership Debate

Climate policy actions are important at all levels – global, regional, national and local. The UNFCCC's global commitments such as the *Kyoto Protocol* and the *Paris Agreement* promote ambition and agreement and monitor action (Chapter 1). These frameworks set out the strategies to be followed by nation states guided by the principle of common but differentiated responsibilities aimed at ensuring equitable sustainable development (Table 3.1; Pauw et al., 2019). However intermediate, regional level efforts are critical – hence our focus on EA&P and the need to reflect upon regional climate leadership within it. Debating the meaning of climate leadership in a regional context expands the focus of inquiry beyond the achievements of nation states, for example, by harmonising climate action with broader concerns with relative levels of economic development, wealth generation and vulnerability reduction within the region (Chapter 5; CSIRO, 2006; McGee & Taplin, 2008). Leadership by developed nations is also considered more wholistically, in terms of domestic *and* regional climate efforts.

Within EA&P, for example, the developed countries of Singapore, Japan, the Republic of Korea, Australia and New Zealand have the means, ability and experience to assist developing nations within their region (CSIRO, 2006; Habib, 2011) as Germany and previously the UK have had in the EU. And, as Chapter 2 argues, EA&P is a significant region in terms of its size, diversity and global impact, and one which will determine whether the goals of the *Paris Agreement* are met. Dent (2012) describes East Asia as one of the world's most important regions geopolitically and geoeconomically with influence that has increased dramatically in recent decades. The capability and influence of the region are growing rapidly along with its economic, political and sociocultural growth, and as China emerges as a global actor and climate leader (Chapter 7). The capacity of the region to respond to climate change will now be key to securing internationally agreed global emission reduction targets and to averting dangerous climate change.

In this chapter, we draw upon governance and climate leadership concepts to distil our regional leadership normative criteria (Table 3.3). Neither literature is specific to our task, but elements of both are useful. For example, the climate leadership literature has identified pioneers and leaders as important drivers for climate change innovations, including in the effective implementation of policies, instruments and approaches. The climate governance literature emphasises the need to build trust, relationships and regimes at multiple levels and in a polycentric fashion which is useful to consider in the regional

Table 3.1 Common but differentiated responsibility

Leadership	*Country type*	*Rationale*	*Whose responsibility*	*What is required to reduce emissions?*	*Leadership challenges*	*Climate leadership?*
Climate leadership typically expected	Developed countries	Developed countries should fulfil their agreed international commitments in line with the 2030 and 2050 emission reduction goals of the *Paris Agt**	In UNFCCC agreements, developed countries have accepted that it is their responsibility to act first on climate change by reducing emissions	Developed countries need to lead reduction efforts based on principles of common but differentiated responsibility and also relative to country capabilities And they need to act in acknowledgement of the collective developed country responsibility for acting first on GHGs	Successful leadership requires both transitioning to low carbon economies and supporting developing countries to do the same	Developed countries have not met expectations Mitigation, adaptation and financial help to lesser developed and developing countries are inadequate
Climate leadership not typically expected	Lesser developed & developing countries	Developed countries should fulfil their *Paris Agt* commitments before developing countries increase their ambition	Developing countries have accepted the responsibility to act, but only as their capacity, capabilities, development trajectories and finances allow	Developing countries have called for the agreed principles of 'common but differentiated responsibilities' to be respected and acted upon Lesser developed & developing countries have pledged to reduce emissions but expect the support of developed countries to do so	Successful leadership requires a managed and assisted transition to low carbon and societal economic development	Leading actions are hampered because climate financial support to developing countries is entirely inadequate

Sources: Hurri (2023a); UNEP (2023); UNFCCC (1992).
Note: *Paris Agreement.

context. Whilst the climate leadership literature invariably focuses on nation states, except for literature about the EU's efforts as a regional actor, the climate governance literature advances multifaceted, multi-levelled connected notions of action (Egerough-Adindu, 2022). Without specifically addressing climate leadership or geographic belonging, it proposes various useful polycentric, regime and building-block approaches to bridging levels of climate action (Stewart et al., 2013). Although a wide range of actors have been identified in climate change governance, studies of climate leadership rarely look beyond nation states to other actors (e.g. international organisations, NGOs, corporations, and cities) as climate leaders.

The climate leadership literature is less useful where omits consideration of each of the differing levels of climate governance, (e.g. local, national, regional, supranational and international), and the significance of regional level policy efforts. The proliferation of a wide range of competing and steadily fracturing theoretical definitions of what constitutes a leader and/or pioneer in climate change governance is also less useful for our criteria-based comparative policy research in the EA&P, although it is useful theory building (Chapter 8). Our approach also has its own limitations. Our leadership criteria are a normative representation of the sorts of actions by which to identify a climate leader in its own region, but they more readily apply to developed rather than developing countries. We employ these criteria in assessing the efforts of select EA&P countries (Table 3.2) in Chapter 4 and find, for example, that they are less applicable to the Pacific Islands and small island developing states (SIDS). We also encounter data gaps and the limitations of empirically informed comparative attempts to ascribe regional climate leadership which leads us to a focus on available, rather than best indicative, criteria. We also appreciate that the Climate Change Performance Index and the Climate Action Tracker have already established that none of the EA&P country efforts align with the goals of the *Paris Agreement*, let alone lead within the region.

Our approach is, however, solution oriented for highlighting specific leading climate policy actions that could be taken in the regional context, and that are credible, effective, responsible, coherent, committed to regional capacity building and cognisant of differing country capacity. Other approaches are complementary to our focus on supportive action within the region. Dent (2008, 2012), for example, conceives of leadership in East Asia as encompassing the actions of a nation state that benefit the region, which is a notion that

Table 3.2 Select countries in East Asia and the Pacific

Category	*Country*
Developed countries	*Australia, Japan, New Zealand and the Republic of Korea*
Lesser-developed countries	*Malaysia and Taiwan*
Developing countries	*China, Indonesia, Thailand and Viet Nam*

is well suited to climate leadership within and by the Pacific Island community. He defines regional leadership as a collective endeavour which includes leading actions in regional community building, resolving collective action problems and championing and representing the interests of the region in the wider global community. Hurri (2023a) also conceives of climate leadership as a collective, if not as a specifically regional, endeavour. She identifies several key climate leadership discourses, in terms of 'responsibilities' (developed country or collective responsibility), 'problematisations' (including social and developmental questions or not) and 'solutions' (in transitioning to low carbon development). Finally, Wurzel et al. (2021) focus on structural, entrepreneurial, cognitive and exemplary climate leadership (Chapters 7 and 8), in examining climate policy efforts across the globe, and in multi-level and polycentric contexts.

Despite the analytic problems of criteria-based analysis, we maintain that what is missing in the climate leadership literature is a systematic conceptualisation and analytical review of potential regional building blocks of climate change actions and of nation states acting within the regional context to build climate policy and resilience capacity. Regional climate leaders should, we argue, act decisively in terms of their domestic mitigation and adaptation efforts but should also directly or indirectly assist and influence other nations within their region, particularly lesser-developed and developing nations, to do the same. The agreed principles of common but differentiated climate policy efforts could therefore be usefully considered in the 'regional climate leadership' context within EA&P (Table 3.1) as developed countries in the region assisting lesser-developed and developing countries to exit fossil fuels, for example by financing mitigation and adaptation and helping to design transition pathways to low carbon economies. As Chapter 5 makes clear, lesser-developed and developing EA&P countries lack the capacity to act on climate change with the urgency that is required to avert its dangerous impacts, and the onus is surely upon developed countries within the region to help.

Such help is urgently needed. Previous chapters have detailed the extreme vulnerability of EA&P to climate change, as one of the world's most disaster prone and heavily impacted regions, and in Chapter 5 we detail preparedness, adaptation and policy resilience challenges within the region. The EA&P suffered 481 recorded natural disasters in the form of floods, storms and earthquakes between 2015 and 2020 (World Bank, 2023b). Over the last 50 years, it has witnessed increases by a factor of 11 in floods, by a factor of 4 in storms, by a factor of 2.4 in droughts and by a factor of 5 in landslides (UNICEF, 2023). The scale of natural disasters and the added impact of climate change, combined with population growth, urbanisation and concentration of populations in disaster prone regions, all heighten the potential massive loss of life, and damage to property and infrastructure, and greatly increase risk (Thomas et al., 2013). Climate change is therefore an existential threat in EA&P and the assistance, including climate finance, of developed countries, especially for the small island developing states of the

Pacific, is crucial (Chapter 5). Moreover, lack of preparedness, poverty and regional tensions each exacerbate the social, economic and environmental impacts of climate change and natural disasters, as does the failure to learn from previous disasters and to transfer learning across the region. There is an equal, if differentiated, challenge for EA&P nations in building resilience against disaster and climate change impacts, which, we argue, requires climate leadership from developed nations.

Considering Regional Climate Leadership

We have argued that it is appropriate for EA&P countries – particularly those with a good history of cooperation – to develop climate action 'neighbourhoods' with a strong awareness of, and concern for, the climate change efforts of, and impacts upon, their neighbouring countries (Crowley & Nakamura, 2015, 2017). In Chapter 1, we profile the countries within the EA&P region, as defined by the US Bureau of East Asian and Pacific Affairs, in terms of their population, gross domestic product, per capita emissions, % of global emissions and emissions change from 1990 to 2019 (Table 1.1). The selection of EA&P countries that appears most economically capable, in terms of their GDP, of reducing their GHG emissions and of assisting other countries in the region are the developed countries Australia, Japan, New Zealand and the Republic of Korea. The countries next most economically capable are Malaysia and ROC Taiwan (Taiwan); however, we also include, for contrast, the developing countries of Indonesia, Thailand and Viet Nam. China is also included as a developing country because it is the world's largest GHG emitter and second largest economy, with enormous leadership capacity which we consider in Chapter 7. These developed, lesser-developed and developing countries feature in more detailed climate leadership analysis against our criteria in Chapter 4, with the expectation that developed country efforts would be leading in the region (Table 3.2).

Previously, we distilled comparative regional climate leadership criteria from the climate leadership and governance literature, namely: 'credibility and effectiveness' and 'responsibility and coherence' (Crowley & Nakamura, 2015, 2017), to which we have added here 'capability and differentiation'. These criteria we have made applicable to nation state action, and to nation state action with the intent of building regional climate capacity and resilience, within 'neighbourly' contexts, and often as bilateral or multilateral, polycentric efforts. Countries in regional proximity should, we argue, act with an awareness of, and a concern for, climate change impacts upon their immediate geographic neighbourhood, and more generally within the region. The first criteria, credibility and effectiveness, are generic to climate leading countries which will be committed to action and will act legitimately in implementing it. The second criteria, responsibility and coherence, are specific to neighbourly and regional climate action and support within the EA&P context, whereby climate leading countries will accept the need for such action

Table 3.3 Criteria for regional climate leadership

Criteria	*Sub-criteria*
Credibility and effectiveness	• making a commitment to act • setting domestic targets • adopting effective instrumentation • not relying on offsetting • transitioning to renewable energy
Responsibility and coherence	• accepting regional responsibility • cooperating on regional climate action • negotiating with major regional emitters • acknowledging humanitarian need
Capabilities and differentiation	• recognising differing country status • recognising differing country wealth • providing climate finance • committing to capacity building • supporting technology transfer

and for it to be coherently undertaken. The third set of criteria, capabilities and differentiation, would see climate leading countries acknowledge and act upon the differing capacities, capabilities and expectations of developed and developing countries within the region. Our leadership criteria are listed in Table 3.3.

Defining Credibility and Effectiveness

Credibility is important for any actor or nation state that aspires to be an effective leader (Underdal, 1994; Grojean et al., 2004). Credibility in a broader sense also refers to a country's 'good standing', whose ambitious actions in implementing climate policy will be believed and emulated by 'follower' countries, because only 'genuine and substantial leadership abatement will motivate other countries' (Schwerhoff & Kornek, 2018, p. 509). Konidari and Mavrakis (2007) adopt credibility as an evaluative criterion to assess the accuracy and consistency of climate actions in terms of verifying the data necessary for the implementation, promotion and steering of national compliance efforts. Effectiveness as a criterion serves a similar purpose (Keohane & Victor, 2011). In terms of regional climate action and leadership, we are concerned about whether developed countries have credible climate policies with effective mitigation and adaptation actions and are thus in the position to advocate, negotiate for and facilitate the same within the EA&P. We see the following elements as crucial to credibility and effectiveness. The academic and practitioner literature all suggests that without a bona fide commitment (in terms of climate target setting, effective instrumentation and a transition to renewable energy) from developed nations, dangerous climate change will not be averted (Averchenkova & Bassi, 2016). The NDC process now provides for a systematic detailing, monitoring and reviewing of nation state commitments (UNESCAP, 2023a).

Making a commitment to act. Commitment to the UNFCCC's regimes and agreements, and to their goals, actions, timetables, participation, institutional arrangements, and provisions for verification, reporting and compliance (IPCC, 2014), is a sign of climate leadership, sends a strong signal to other countries and promotes regional confidence in a climate leading country.

Setting domestic targets. This is critical to inspiring other sectors, levels of government and countries to follow with actions of their own (Eckersley, 2012, 2020; Tobin, 2017), and to motivating global achievements (Andresen & Agrawala, 2002). Ambitious, effective target setting, with credible baselines, is critical for a climate leader in a region such as EA&P.

Adopting effective instrumentation. Climate policies at different levels of governments must be capable of meeting their targets and achieving outcomes (Hovi et al., 2009; Keohane & Victor, 2011). A climate leading country will achieve its national climate policy targets through the adoption of effective instrumentation including national and sectoral carbon pricing policies.

Not relying on offsetting. Credible climate action reduces emissions without over reliance upon flexibility and offsetting mechanisms either domestically or in other countries (Fejka, 2012). The UNFCCC prescribes two climate policy leadership practices: comprehensively covering all GHG emissions; and placing limitations and conditions upon any offsetting (UNFCCC, 2020).

Transitioning to renewable energy. A climate leader will be transitioning away from GHG producing processes and energy systems, towards a low carbon future, and developing new technologies and creating new capacities and jobs (IPCC, 2022b). A regional climate leader assists other nations within the region with making such a transition (Green & Finighan, 2012).

Defining Responsibility and Coherence

Responsibility as a criterion of climate leading behaviour refers to acceptance by a climate leading country that the principles of distributional fairness, burden sharing and common but differentiated responsibility should guide climate policy practice (Torvanger & Ringuis, 2002; Maltais, 2014). Accepting this responsibility and the internationally established notion that developed states should take on the largest emission reducing burdens and act first is, we argue, the key to demonstrating regional climate leadership and is, indeed, expected from the more capable, developed nation states. Any policy action should however be coherent and complementary to a country's policy settings. Coherence is therefore an important evaluative criterion for climate actions and emphasises that these should be compatible and mutually reinforcing, not undermining other climate actions (Keohane & Victor, 2011). Coherence in regional climate leadership, we argue, also refers to policies and actions from developed countries being compatible with the best interests of, or doing no harm to, the broader region. This is important in evaluating the role of fossil fuel expansion in the EA&P region which is undercutting and delaying the uptake of renewable energy. We have adopted these criteria because they infer thinking of regional neighbours and acting beyond the national context and could influence developing nations to act if exhibited by a developed country.

Accepting differentiated responsibility. Climate leaders accept differentiated responsibility, in the regional context by observing the UNFCCC principles of equity between nations and common but differentiated responsibilities, with developed nations expected to lead on climate action (Article 3(1) of the UNFCCC; Article 4.4 *Paris Agreement*). By modelling differentiated responsibility, regional leaders assist with broader global efforts to have developed nations elsewhere do the same.

Cooperating on regional climate action. A leading climate country projects an awareness of its geographic neighbourhood and takes actions to build regional climate action (mitigation *and* adaptation) capacity. The aim is minimising harm from the impacts of climate change, and building resilience and coherent climate policy capacity in a regionally specific sense which again can provide a building block for global action (CSIRO, 2006; Keohane & Victor, 2011).

Negotiating with major regional emitters. A regional climate leader negotiates with and assists major emitters in the EA&P region (such as China, India and Indonesia) in their efforts to reduce their emissions, to lessen their reliance upon harmful fossil fuels and to protect the region. This will involve accepting responsibility and working within existing and new partnerships, and with regional actors, to achieve transformation (OECD & IEA, 2015; Chapter 2; Table 2.1).

Acknowledging humanitarian need. A regional climate leader recognises the unequal burden of climate impacts and is committed to minimising the harm of climate impacts upon regional neighbours, as well as to meeting their humanitarian, loss and damage, needs (Schipper & Pelling, 2006). Regional mitigation and adaptation policies would integrate with EA&P development aims (UNFCCC, 1992, Article 3.4) on the basis that developing countries can least afford to act.

Defining Capability and Differentiation

Capability is often referred to in the context of 'adaptive capacity'. It is a notion that resonates with our criteria of responsibility and coherence and emphasises the ability of nation state systems to respond to the consequences of climate change damage, to adapt, to reorganise and, if applicable, to take advantage of any opportunities (IPCC, 2014). In terms of the ingredients of successful climate change actions, various capabilities have been identified such as 'political stability' (Balcazar & Kennard, 2023), 'institutional capability' (Brown et al., 2010) and 'production and technology capability' (Schive, 1990). It is also widely acknowledged that the countries that are the most vulnerable to the impacts of climate change are often those least able to finance the actions necessary to act on those impacts (WHO, 2023). Financial capacity is therefore critical for effective action to achieving the goals of the *Paris Agreement* (ODI & HBS, 2022). Capability in the context of regional climate leadership highlights the vastly differing country status and wealth in EA&P, the recognition of which is critical to regional climate leadership. Accepting and acting upon the principle of differentiation to assist developing countries build their capacity to act upon mitigation, adaptation and transition to a low carbon future is a climate leading response.

Building upon the internationally accepted differentiation principle are the notions that fairness, equity, and a just transition for developing countries should be endemic to international climate change agreements (Cazorla & Toman, 2000; Fitzgerald, 2022). The *Paris Agreement*, for example, emphasises the financial impacts of loss and damage from the impacts of climate change, and the need for financial restitution to impacted developing countries from the wealthier developed nations with the historical responsibility for climate change (UNFCCC, 2015, Article 8.3; Chapters 2 and 5; Carbon Brief 2015). Acting upon differentiation by a potential regional climate leader can be examined, we argue, by considering its levels of climate finance to the region (OECD 2022a), as well as its funding of climate technology transfer and of adaptive climate capacity and resilience building in developing countries (Pauw et al., 2019). Given the rate at which the emissions of developing countries in EA&P are accelerating as these countries develop, investments by developed countries in building the technical and adaptive capacities of less capable countries in their region will ensure that the goals of the *Paris Agreement* remain within reach.

Regional climate leaders would acknowledge country capability and differentiation principles by ***recognising differing country status***; ***recognising differing country wealth***; ***providing climate finance***; ***committing to capacity building***; and ***supporting technology transfer***.

Recognising differing country status. Developed countries will have leading economic status within a region, although in EA&P China is an exception as a developing country with the status of regional economic leader. A regional climate leader will act with an awareness of differing country status, typically as a developed country providing financial support for capacity building to developing countries, or as a developing country, or indeed as developing regions such as ASEAN and the PIC, actively seeking to elicit such support. Lacking developed country status is a significant impediment to effective action on climate change and to meeting the climate goals established by the *Paris Agreement* and a climate leader will acknowledge and act to address this.

Recognising differing country wealth. Developed country status implies that not only is a country historically responsible for climate change unlike developing countries but also it has the wealth to transition to a low carbon economy and to build resilience against climate impacts. Country wealth is a measure of structural power, which can leverage structural climate leadership (Wurzel et al., 2021) in terms of leading domestic actions, and leading actions in international and regional arenas. A regional climate leading country will recognise that developing countries have less structural power and therefore less capacity to act on climate change and will offer its support to help build the capacity to transition, to adapt to climate impacts and to become resilient.

Providing climate finance. A regional climate leader will therefore provide climate finance to lesser-developed and developing countries. This finance should not become a debt trap for lesser-developed and developing countries and should cause no economic, social or environmental harm but afford the ability to address the causes and effects of climate change. Low-income countries are more vulnerable to current climate variability and future climate change than rich developed countries, including those in EA&P (World Bank, 2024a). A regional climate leader will provide finance recognising that developing countries are least able to develop climate policy and transitioning responses and to fund climate change 'loss and damage' (OECD, 2022a; WHO, 2023).

Committing to capacity building. Article 11.3 of the *Paris Agreement* includes as a priority that all parties should cooperate to enhance the capacity of developing countries, with developed countries required to support capacity building actions in developing countries (Khan, 2017; Pauw et al., 2019). Capacity building plans are to be detailed in country NDCs (described in Chapter 1) which are submitted, updated and reported on to the UNFCC. Indeed, developing countries have emphasised a need for capacity building support from developed countries as a condition for implementing their own NDCs (Khan et al., 2020). Regional leaders should detail, in their NDCs, plans to undertake capacity building actions within both their region and developing countries.

Supporting technology transfer. Developing and lesser-developed countries lack the capability of developed countries when it comes to investment in, and the development and manufacture of, the technology that is required for transitioning to a low carbon economy. Developed countries must address technology transfer in their NDCs consistent with Article 10.4 of the *Paris Agreement* (UNFCCC, 2015; Pauw et al., 2019). Regional climate leaders will therefore assist with the development/uptake of low carbon technology by their developing and lesser-developed country regional neighbours, particularly the large emitters and those countries that are still developing but are highly fossil fuel dependent, to ensure that this transition is underway.

Our regional climate leadership criteria, their rationale and their indicators are normatively depicted in Table 3.4.

Table 3.4 Assessing regional climate leadership

Criteria	*Rationale*	*Sub-criteria*	*Indicators*
Credibility & Effectiveness	Regional climate leaders have credible climate policies with effective mitigation and adaptation actions	*Making a commitment to act*	International commitments made (e.g. *Kyoto Protocol*, and *post-Kyoto/Paris* regimes) consistent with UNFCCC goals
		Setting domestic targets	Scope of the GHG emissions reduction target is ambitious and credible
			Target's legislative strength is ambitious and credible
			Emission reduction targets are economy wide
		Adopting effective instrumentation	Economy-wide/sectoral carbon pricing policies are in place that avoid fossil fuel subsidies or unreasonable exemptions
			Scope of the pricing mechanism is transformative
		Not overly relying on offsetting	Committed to meeting targets based on the country's own initiatives to reduce emissions, without overly relying upon flexibility mechanisms (e.g. LULUCF accounting rules)
		Transitioning to renewable energy	Adopted legislative instruments for renewable energy
			Bona fide transitioning to low carbon energy (not relying on nuclear power plants and/or the fossil fuel industry)
Coherence & Responsibility	Climate leaders cognisant of regional neighbours and of acting beyond the national context and inspiring other countries	*Accepting regional responsibility*	Committed to responsibility and to developing nations acting first on climate mitigation and adaptation at the global and regional levels, and to assisting other nations.
		Cooperating on regional climate action	Committed to cooperative policy actions in the region
			Cooperation on polycentric climate action with the region
		Negotiating with major regional emitters	Negotiations on mitigation with the significant emitting countries in the region (e.g. China, India, and Indonesia)
		Acknowledging humanitarian need	Provided the means for meeting the humanitarian needs of threatened developing countries within the region
Capability & Differentiation	Leading EA&P countries recognise the differing capabilities of countries within the region to act on climate change and their responsibility to commit to help build capacity in lesser developed and developing countries	*Recognising differing country status*	Has acknowledged leading and lagging economic status in the region (e.g. developed, lesser developed or developing).
		Recognising differing country wealth	Has provided for adaptive climate capacity building in its own country and for threatened regional neighbours
		Providing climate finance	Has provided climate financial support for countries in the region (i.e. financing climate mitigation and adaptation)
		Committing to capacity building	NDC details specific support for the capacity building of developing and least developed countries in the region
		Supporting technology transfer	NDC details specific technology transfer, including financial support for technology uptake, from developed countries to developing and least developed countries in the region

Sources: IPCC (2014; OECD (2022b); UNFCCC (1992, 2015, 2019).

Conclusions

We have defined regional climate leadership broadly as 'taking, while encouraging and assisting others to take, effective actions to reduce emissions and to adapt to climate change both domestically and within the region'. We have argued that the systematic investigation of such leadership is missing beyond the context of the EU and that the notion of regional climate leadership has potential relevance to the EA&P. Given the fractured institutional context in EA&P, and in the absence of a unifying institutional environment such as the EU's, regional climate leadership could be depicted as leading policy efforts taken domestically and within the EA&P context. This polycentric setting would be conducive to promoting, transferring and accelerating climate policy learning of mutual benefit to neighbouring countries, facilitated by bilateral and regional forums and agreements. We have devised a framework for the comparative analysis of a selection of developed, lesser-developed and developing EA&P countries in Chapter 4 to identify leading climate policies and actions, and potentially leading countries. Our regional climate leadership criteria 'credibility, effectiveness, responsibility, coherence, capability, and differentiation' broaden the notion of a climate leader to one which is regionally focused and collaboratively active in building the capacity of other nation states within the EA&P.

Our approach therefore aims to probe domestic climate efforts, but in the context of the need, we argue, to show regional climate leadership and to act upon the principles of differentiation by offering mitigation and adaptation assistance to less capable countries within the region. In climate governance terms, the application of our framework offers to bridge the global–domestic level of climate policy analysis with an intermediate focus on the regional level of analysis while also addressing the challenges of developed, lesser-developed and developing nation states' climate policy relations that have been highlighted as significant by international climate governance processes. Our approach suggests that climate leadership in regional, polycentric, multi-levelled circumstances is worthy of further attention for adding 'collaborator' to the language of 'climate leader' and 'climate laggard' and for capturing a collaborative style of leadership that is relevant in settings such as the EA&P. Next, we will establish the extent to which EA&P countries with the greatest capacity to act on climate change domestically, and within their region, namely the developed countries of Australia, Japan, New Zealand and the Republic of Korea, and the lesser-developed countries of Malaysia, Taiwan, China, Indonesia, Thailand and Viet Nam, and in the Pacific, have in fact done so, and if any are acting as regional climate leaders.

4 Policy Action, Efficacy and Climate Leadership in EA&P

Introduction

In Chapter 1 we saw that none of the countries in the East Asia and Pacific region (EA&P) with the greatest capacity to act on climate change, in terms of GDP (Table 1.2), feature well on the Climate Change Performance Index (2024). China, Malaysia and Korea rank very lowly at 51st, 59th and 61st as may be expected, but the developed countries in the region from which leadership could be expected rank extremely poorly – New Zealand (low), Australia (low) and Japan (very low) ranking 34th, 50th and 58th of 63 countries. By contrast, one of the least developed, least capable countries, Thailand, ranked moderately well at 25th and, despite its lack of capacity, Viet nam is the highest ranked of these countries as an emerging climate leader in the EA&P ranked 27th and so is added here to our analysis. Each of these countries has its own climate change challenges. China, for example, is still a developing country with a low carbon footprint, but its emissions have burgeoned by more than 200% since 1990, and it is now the world's largest emitter of greenhouse gas (GHG) emissions (Table 1.2; Chapter 7).

The picture drawn by CAT is even bleaker. The best ranked of this selection of countries, Australia and Japan, have 'insufficient' climate performance rankings. New Zealand, South Korea and China have 'highly insufficient' rankings for lacking the required policies to meet the *Paris* targets. As developing countries in the EA&P, Thailand, Indonesia and Viet Nam have 'critically insufficient' rankings (CAT, 2024). The CAT and CCPI rankings alone do give a good indication of leading and lagging efforts and provide the breakdown of these across a range of criteria: policy, GHG emissions, energy use and renewables (CCPI analysis); and policies, actions, NDC targets and climate finance (CAT analysis). Our analysis here provides an additional normative overlay to view criteria for defining climate leadership in a regional context, not only in terms of leading domestic actions but also actions which build climate policy capacity with the expectation that developed nations will assist lesser-developed and developing nations in the regional context.

DOI: 10.4324/9781003170389-4

Assessing Credibility and Effectiveness

Developed Countries – Australia, Japan, New Zealand and South Korea

We have argued that credibility and effectiveness are key criteria for measuring climate leadership within the EA&P region. We can measure this, at least in terms of intentions, by looking at: making a commitment to act, setting domestic targets, adapting effective instrumentation, not relying on offsetting and transitioning to renewable energy (Table 3.3). Australia, Japan, New Zealand and South Korea are committed to the *Paris Agreement.* Japan, New Zealand and South Korea's NDC targets aim to limit temperature increases to 1.5 °C above pre-industrial levels, with only Australia committed to the lesser target of 2 °C. Each country has legislated a net zero emission target by 2050 (Table 4.1), but these are not compatible with the 1.5 °C goal (CCPI, 2024). However, target setting loses relevance where these countries' policies and actions to 2030 are inconsistent with the 1.5 °C *Paris Agreement* goal. If all countries replicated this failure to meet their emission targets, then global warming could reach 3 °C to 4 °C by 2050 (CAT, 2024). Australia, Japan, New Zealand and South Korea thus lack credible, effective emission reducing targets, policies and instrumentation.

In terms of offsetting, Australia has negotiated to increase its net GHG emissions since the 1997 *Kyoto Protocol.* Its emissions reduction strategy relies on domestic and international offsets (i.e. LULUCF) (by c. 10–20%). Its emissions reduction target is 43% below 2005 levels by 2030, including LULUCF. This translates to a 24% reduction below 2005 levels by 2030, excluding LULUCF. However, for a 1.5 °C compatible NDC target against modelled domestic pathways, Australia needs reductions by 2030 of at least 61% with LULUCF and 44% without LULUCF. This is a historic problem (Crowley, 2007). Similarly, New Zealand and Japan rely heavily on carbon offsetting for emissions reductions (from both LULUCF sinks and international credits) to meet their 2030 targets. These are not credible nor effective actions and so none of these countries are climate leading for avoiding offsetting. South Korea does not rely upon offset mechanisms. However, its climate policies and long-term strategy are 'highly insufficient' not only in meeting their targets but also in terms of compatibility with the 1.5 °C goal (CAT, 2024).

In terms of credible, effective mechanisms to assist the transition to renewable energy, New Zealand and South Korea have implemented nationwide carbon pricing, although we note carbon pricing's many general and country specific shortcomings in Chapter 6. In Japan, only two sub-regional governments (the Tokyo Metropolitan Government and the Saitama prefectural government) have introduced emissions trading schemes (ETSs), although the national carbon pricing scheme is set to begin in 2026. An incoming Conservative government repealed Australia's newly implemented carbon pricing scheme in 2014 (Crowley, 2017), although the current Labour government will revive the previously ineffective Safeguard Mechanism

Table 4.1 CAT/CPPI – Climate policy credibility & effectiveness from developed countries in EA&P?

Country	*2030 goal*	*2050 goal*	*RE target*	*Fossil fuels*	*Carbon price?*	*Offsetting?*	*Other credibility issues?*
Australia	43% vs. 2005 levels	Net zero	82% by 2030	No phase out. Indeed, strong policy support for fossil fuels in general and as a key export industry	Abandoned in 2014. Potential for the questionable Safeguard mechanism to work as a quasi-carbon price	Australia has revised its LULUFC sequestration date to significantly ease the path to meeting its climate targets	Yes – the credibility of both Australia's Safeguard Mechanism and its offset processes in general have been questioned
Japan	46% vs. 2013 l evels	Net zero. Also carbon neutral by 2050	No	No phase out – Japan will keep using coal till 2050 whilst trying to curb emissions from coal-fired power plants through new technology	No economy wide carbon price, although one is being suggested. Smaller schemes exist in Tokyo and in the Saitama prefecture	International offsets will be used as part of a plan to cut emissions by 46% between 2013 and 2030. Offsets are claimed for carbon dioxide absorbed by the country's forests	Yes – Japan doesn't include LULUFC in its baseline year but does include it in its target year. This practice undermines the UNFCCC accounting system

(*Continued*)

Table 4.1 (Continued)

Country	*2030 goal*	*2050 goal*	*RE target*	*Fossil fuels*	*Carbon price?*	*Offsetting?*	*Other credibility issues?*
South Korea	40% vs. 2018 levels	Net zero. Also carbon neutral by 2050	21.6% by 2030 Share of RE in the energy mix the lowest in the OECD	No. Retired coal power plants will be replaced by gas plants. Public financing of oil and gas. More nuclear is planned	Nationwide carbon pricing. Coverage of the Korea Energy Trading System (K-ETS) was increased from 68% of national GHG emissions to 73.5% in 2021	Relies on international offset credits or reserves the right to use them to meet its targets. Domestic mitigation efforts including LULUCF to achieve its targets	Yes – a change of government has seen a weakening of climate ambition and a slowing of progress against targets, with national interests taking precedence
New Zealand	30–50% below gross 2005 levels inc. LULUCF	Net zero	50% by 2035 – but hoping for 100% by 2030	No – most of NZ's energy is supplied by fossil fuels. But NZ signed an international pledge in 2021 to phase them out	Nationwide carbon pricing does not include agriculture – 40% of NZ's emissions	Relies on international offset credits or reserves the right to use them to meet net zero. Reliance on LULUCF to meet NZ's NDC target decreased to 26% from 35% since 2022	Updated NDC based on two misleading accounting methods that more than halve NZ's effective reduction in net emissions (including LULUCF emissions) to only 22% below 2005 levels by 2030

Sources: CAT (2024); CCPI (2024).

Note: RE – renewable energy.

as a quasi-carbon pricing tool (see Chapter 6; Table 4.2). All four countries rely heavily on fossil fuels (Table 1.2) including for the future development of low-emissions technology with no clear plans to phase out coal, curb fossil fuel exports or hold heavy polluters accountable (CAT, 2024; Chapter 2). CAT concludes that none of these developed, lesser-developed or developing countries have a clear plan for transitioning to renewable energy despite our Table 4.2 findings. None are therefore climate leaders in the region in terms of adopting credible, effective mechanisms to assist the transition to renewable energy.

Lesser-Developed Countries – Malaysia and ROC Taiwan (Taiwan)

Malaysia signed the *Paris Agreement* but is not required to report its NDC long- or medium-term emission reduction targets. Its net zero GHG emissions 2050 target is not yet legislated. Taiwan is not recognised as a member of the United Nations, so is excluded from UNFCCC processes and agreements (Mabon, 2021), and is not a signatory either to the *Kyoto Protocol* or the *Paris Agreement*. However, it has achieved observer status, is an active participant at the COP international annual meetings and has voluntarily adopted the *Paris Agreement's* 1.5 °C target and passed a *Climate Change Response Act* with a legally binding target of net zero GHG emissions by 2050. In terms of policy instrumentation, Malaysia is at the exploratory stage of carbon pricing and has established a carbon exchange (Chapter 6) although it continues to subsidise carbon emissions. It has not registered offsetting projects with the UNFCCC's carbon offset platform. CCPI (2024) acknowledges that Malaysia's *National Energy Transition Roadmap* (NETR) does establish a low carbon pathway for the energy sector. But there is insufficient information to date for making any assessment of the credibility or effectiveness of these measures, and the country heavily depends on the oil and gas industries which are subsidised at many levels, including blanket subsidies for transport fuel (CAT, 2024; CCPI, 2024). These lesser-developed countries are not expected to be climate leaders, although Taiwan is exhibiting leadership tendencies on several criteria.

Taiwan has voluntarily adopted both an ETS and a CT as a means of achieving its voluntary net zero by 2050 target, with the Taiwan Carbon Solution Exchange becoming operational towards the end of 2023. It is in active dialogue with the Asian region, the U.S and Europe on developing its carbon market bilaterally, regionally, and internationally and is seeking to overcome the obstacle that its lack of formal 'country status' presents (there are country status limitations to international carbon market 'parties' in formal international treaties) (Mehling et al., 2013). Taiwan has not registered projects with the UNFCCC's carbon offset platform; however, it has established the GHG Offset Project as 'a mechanism for issuing credit to encouraging implementation of voluntary GHG emission reduction by entity' although there is, thus far, insufficient information to determine the

Table 4.2 Transition planning in EA&P

Australia	2023 – Australia announced a collaborative approach with states and territories under the new *National Energy Transformation Partnership* but has yet to define a framework and policies under its *Powering Australia* plan for transitioning to net zero emissions by 2050
Japan	2021 – Japan's carbon intensive energy supply is one of the highest of IEA members. Its enhanced RE and energy efficiency plans will need to achieve rapid and steep emissions reductions to reach its recently announced ambition of achieving carbon neutrality by 2050
South Korea	2020 – Korea is highly reliant upon fossil fuels/imported energy. Its 'green new' carbon neutrality by 2050 will be achieved by boosting renewable energy, phasing out coal, enhancing energy efficiency & developing its hydrogen industry, fostering green growth and innovation
New Zealand	2023 – over 80% of electricity is from renewables. But NZ must invest in decarbonising all other sectors, the transport and industrial sectors in particular which are reliant upon oil and fossil fuels, to achieve net zero emissions by 2050. A long-term energy strategy is required
Malaysia	2024 – the *National Energy Transition Roadmap* will see coal plants phased out by 2025; and identifies six energy transition levers including renewable energy, hydrogen, bioenergy, green mobility, energy efficiency and carbon capture, utilisation, and storage
Taiwan	2023 – implementing *Taiwan's Pathway to Net-Zero Emissions in 2050* will see carbon fees reinvested into low carbon initiatives and technologies; and plans adopted for energy security, industry competitiveness, societal resilience, and low carbon, sustainable living
China	2021 – China released *An Energy Sector Roadmap to Carbon Neutrality in China* showing that gaining carbon neutrality fits with its broader RE & development goals (increasing prosperity and innovation-driven growth) with an *Accelerated Transition Scenario*
Indonesia	2022 – Indonesia is the world's fourth-most populous country, seventh-largest economy, twelfth-largest energy consumer and largest coal exporter. It requested the IEA to produce its 2022 *Energy Sector Roadmap to Net Zero Emissions* to guide its decarbonisation efforts
Thailand	2020 – the IEA finds that Thailand's ambitious decarbonisation plans in its 2020 *Power Development Plan* will require a rapid scale-up of clean generation to align power sector development with Thailand's domestic and international climate commitments
Viet Nam	2023 – Viet Nam released its *Power Development Plan VIII* with ambitious renewable energy goals but questions over its implementation. Viet Nam is also wary of accepting loans that exacerbate its debt rather than grants to help implement its plan (PwC, 2023b)

Sources: DIS (2023); IEA (2020b, 2021a, 2021b, 2022a, 2023a, 2023b, 2023c); MOEcon (2024); PwC (2023b).

Note: IEA – International Energy Agency; RE – renewable energy.

credibility or effectiveness of this project (MOE, 2024). It imports nearly all its energy, much of which is heavily subsidised to smooth out energy price volatility, and its energy is consumed 80% by industry and only 20% by the community (Yau & Chen, 2021). Because it is not recognised as a country, Taiwan's climate performance is not measured by CAT or CCPI; however, Taiwan is at least exhibiting leadership tendencies, which we depict as intentional leadership.

Developing Countries – China, Indonesia, Thailand and Viet Nam

Developing countries are not expected to be climate leaders. Indeed, they should be supported in their mitigation and adaptation efforts by developed countries within the region. China, Indonesia, Thailand and Viet Nam signed the *Paris Agreement* as developing countries, but their efforts to date to meet either the 2 °C or 1.5 °C goals in their submitted NDCs would in fact lead to a dangerous 3 °C to 4 °C world if adopted by all countries (CAT, 2024). While Thailand and Viet Nam target net zero GHG emissions by 2050, China aims to peak its emissions by 2030 and to reach 'carbon neutrality' before 2060 (Chapter 7). Indonesia, which is heavily fossil fuel dependent, is committed to meeting its targets – both 'unconditional' (without international assistance) and 'conditional' (with international financial assistance) by 2030. CAT recognises that Thailand has raised its ambitions with its second NDC and with strengthened targets in its recently adopted Long-Term Low Greenhouse Gas Emissions Development Strategy. Viet Nam is a developing country pioneer for at least legislating the net zero target of its *National Strategy for Addressing Climate Change by 2050* (Decision No. 896/QD-TV) (CAT, 2024). However, CCPI (2024) notes that Viet Nam's climate performance is marred by obscure implementation and its political repression and imprisonment of climate activists and experts.

Each of these countries relies heavily on carbon offsetting using reductions from LULUCF sinks and international credits to meet their targets in 2030 which mars their climate policy credibility. However, in terms of effective instrumentation, there are currently nine subnational ETSs and one nationwide ETS in China (Chapter 6). In September 2022 Thailand launched its voluntary carbon credit exchange in a bid to curb its GHG emissions. Viet Nam has legislated a domestic carbon market with an ETS and a CT (Aoki, 2022; Chapter 6) and is in the process of implementing them. However, CAT (2024) observes that these countries rely heavily on low-emissions technology that *may* eventuate in the future and have no plans to phase out coal, curb fossil fuel exports or hold heavy polluters accountable. CAT rates the climate policy performance of these countries as China 'highly insufficient' and Indonesia, Thailand and Viet Nam 'critically insufficient' for their lack of ambitious target setting, long-term strategy with climate policies and measures, and lack of plans to phase out fossil fuels, none of which is compatible with reaching the 1.5 °C target.

Pacific Island Countries

Pacific Island countries are overly reliant upon fossil fuels (Pacific Community, 2022), despite generating negligible emissions (Table 1.2) themselves. They are disproportionately the victims of the accelerating impacts of climate change, committed to the goals of the *Paris Agreement*, and have pledged to achieve carbon neutrality by 2050. In Chapter 5, we consider how the various Pacific regional action plans are building an integrated regional response to risk, vulnerability and the need to transition to low carbon economies. There is no expectation of climate leadership from these countries and no tracking of their efforts and NDCs by the Climate Action Tracker (CAT). As a small developing country, Fiji is, for example, not considered in either CAT or CCPI rankings, but it has submitted an NDC with 2030 targets and aims to reach net zero by 2050. It has also enacted climate change legislation to progress its decarbonisation, emissions trading, adaptation and climate finance mobilisation plans (Fiji, 2020). More broadly, the regional *Framework for Resilient Development in the Pacific 2015–2030* integrates climate action, adaptation, disaster and risk management planning into strategic guidance for national climate planning and NDCs by Pacific Island countries (see Chapter 5). There is scope, beyond our analysis here, for assessing regional climate leadership in the Pacific in terms of our criteria and how they could be reconceived in the Pacific context.

Credibility and Effectiveness?

Measuring credible and effective climate policy actions is an inexact science or exercise. Simply making a global commitment to act, setting domestic targets, adapting effective instrumentation, not relying on offsetting, and transitioning to renewable energy does not imply climate leadership. These are relatively commonly applied criteria for measuring climate policy performance but would only be considered leading actions by CAT and CPPI if they were compatible with meeting the 2030 and 2050 emissions reduction goals of the *Paris Agreement.* On this basis, none of the EA&P countries we have considered are regional climate leaders, certainly not the developed countries; however, we have seen signs of intentional leadership from Taiwan, Viet Nam and Fiji. Australia, Japan, New Zealand and South Korea need to assume greater ambition and leadership in their climate policies and actions by implementing plans to cut, rather than offset, emissions and to exit fossil fuels. As CAT (2024) observes, to be *Paris Agreement*-compatible, coal needs to be phased out by 2030 at the latest, and gas needs to be phased out from the power sector shortly after that. CAT and CPPI analysis shows why these countries are not climate leaders, wedded as they are to fossil fuels, lacking effective carbon pricing, and reliant upon offsetting rather than accelerating emissions reductions (Table 4.1).

Whilst none of the EA&P countries are climate leaders according to our criteria, each of them has, however, recently released plans to transition to net zero by 2050 (Table 4.2).

Assessing Responsibility and Coherence

Developed Countries – Australia, Japan, New Zealand and the South Korea

Responsibility is a widely recognised criterion of effective climate action for its emphasis on distributional fairness, whilst coherence emphasises that a country's actions are mutually reinforcing, not undermining climate policy efforts. We have suggested several criteria by which to ideally measure responsibility, none of which readily draw upon exact or accurate data. These are that a regional climate leader would accept regional responsibility; cooperate on regional climate action; negotiate with major regional emitters; and acknowledge humanitarian needs. We rely on various data sources here and in Chapter 5 and, having done this, understand that self-reporting by countries is not guaranteed to provide a credible source by which to measure climate policy effectiveness. Our efforts suggest the need for more forensic country-based case study analysis beyond our scope here. 'Responsibility' can be assumed as accepted by signing up to and participating in international, regional and bilateral agreements (Table 2.1), but curiously does not feature in NDCs (UNDP, 2021). Although we have no specific criterion for 'coherence', we have seen in reviewing both CAT and CCPI analysis of climate policy performance that an over reliance upon offsetting and upon fossil fuels for domestic energy and for export in EA&P countries, is significantly undermining climate policy actions (Chapter 2; Table 4.1).

Australia, Japan, New Zealand and South Korea have accepted regional and differentiated responsibility as signatories to the *Paris Agreement* and in their role as regional leaders, although this is not mentioned in terms of climate action in their NDCs. They have acknowledged the need to assist developing countries with climate adaptation and resilience building; and to support developing and regional nations with adaptation and vulnerability. Australia, for example, is a strong financial supporter of the countries of the Pacific and their climate resilience and disaster response efforts. While these countries may also provide financial direct aid for carbon reducing technology development, their efforts do not match the assistance that is required (Chapter 5). Of these countries, only Japan is spending on climate loss and damage – pledging US$10 million (UNFCCC, 2024b), and its *Joint Crediting Mechanism* has assisted 29 developing countries jointly to reduce emissions since 2013 (MOFAJ, 2024). Regional engagement on climate change and with the major emitters, essentially China (Table 4.3), but potentially also Korea and Japan, is an ongoing diplomatic exercise carried out in a variety of fora, summits, groupings and dialogues. Bilateral dialogue with China within the region is constrained by security and trade concerns, but there are ample opportunities

Table 4.3 Responsibility – engaging with EA&P major regional emitter – China

China	Engaged in talks, for example, as detailed below (see also regional engagement in Table 4.3
Australia	2023 – commenced talks on 'climate change, energy, and environment, including through recommencing the bilateral climate change and energy dialogues, and commencing technical cooperation on soil carbon testing and climate-smart agriculture practices'
Japan	2023 – 24th Tripartite Environment Ministers Meeting (TEMM) between Japan, South Korea and China agreed to strengthen measures to address climate change; accelerate the transformation towards a net zero economy; and improve environmental governance
South Korea	2023 – Grossman (2023) notes that the late 2023 resumption of the TEMM dialogue between Japan, South Korea and China which had been dormant since 2019 is a breakthrough in relations and shows how climate and security dialogue are interrelated
New Zealand	2023 – New Zealand signed a joint comprehensive strategic partnership with China which 'highlighted the importance of concrete action to address climate change and welcomed ongoing productive exchanges in this area, including regular Ministerial Dialogues'
Malaysia	2023 – Malaysia and China celebrate ten years of their China-Malaysia Comprehensive Strategic Partnership, with no specific mention of climate change cooperation, as part of China's Belt and Road initiative
Taiwan	2023 – Taiwan is not recognised as a country by China, but as a legitimate part of Chinese territory. Taiwan has nevertheless established a presence holding side events and bilateral dialogue at COP events on a range of issues including transitioning to a net zero world
Indonesia	2023 – Indonesia is the world's top exporter of coal (Table 2.2). It is in dialogue with China on gaining support for renewable energy via the Belt and Road initiative; and with the USA on a US$20 bill decarbonisation loan under the Just Energy Transition Partnership (JETP)
Thailand	2022 – Thailand and China agreed on mutual cooperation and dialogue across a range of issues including climate change, with China offering support under its Belt and Road initiative, and the statement affirming the ASEAN-China partnership
Viet Nam	2023 – Viet Nam and China have agreed to deepen cooperation on their climate change response within the UNFCCC and *Paris Agreement* frameworks. Viet Nam has agreed to participate in China's Belt and Road Initiative International Green Development Coalition

Sources: Grossman (2023); MOE (2023); MOFA (2022); PMA (2023); PRC (2023a, 2023b); Suroyo & Sulaiman (2023); VLLF (2023).

for multilateral dialogue as depicted in Table 2.1. Again, forensic case study analysis is needed to assess the effectiveness of engagement in advancing climate policy efforts and to be able to establish any leading actions. At a basic level, these four developed countries do accept regional responsibility. They do cooperate and negotiate within the region, and they do acknowledge differentiated responsibility in principle, however with no outstanding shows of climate leadership.

Lesser-Developed Countries – Malaysia and ROC Taiwan

Lesser-developed countries have been drawn more into the global effort to reduce emissions under the *Paris Agreement* even though it is understood that they are less capable of acting and are not historically responsible for global climate change and its impacts. They are not expected to be regional leaders. Lesser-developed countries like Malaysia engage in multilateral dialogue and cooperation on regional climate action through various ASEAN and ASEAN+ groupings well beyond Southeast Asia including with China, Australia, New Zealand and the United States (Table 2.1). Malaysia expects to be a net recipient of international funds to adapt to climate change – for example, for the protection of its forests and biodiversity and to reduce flooding. Taiwan is excluded from the UNFCCC and ASEAN fora, although it is included in APEC. It has limited official opportunities to connect with the global community, for example, to negotiate with major regional emitting countries and industry under the *Paris Agreement* Article 6 global carbon market mechanism that seeks to promote reduced global carbon abatement costs (Chapter 6). It has established a US$10 million Climate Transition Fund for climate change adaptation to assist four Pacific allies, Tuvalu, Nauru, the Marshall Islands and Palau (ROC Taiwan, 2023). Neither Malaysia nor Taiwan aspires to climate leadership, but Taiwan is exhibiting climate leadership tendencies on the global stage, is seeking greater regional engagement and is offering financial support on climate action.

Developing Countries – China, Indonesia, Thailand and Viet Nam

China, Indonesia, Thailand and Viet Nam have been expected to develop economically and socially before acting upon climate change, until their burgeoning emissions caused international concern, particularly with China now the world's largest GHG emitter. These countries have accepted their need to act and have ample opportunities for multilateral climate dialogue, cooperation on regional climate action and negotiation across the region through established channels (Table 2.1). China remains a developing nation and has negotiated internationally on behalf of the interests of the developing countries at COP's international meetings. Chinese governments have argued that, as a developing country, China will not pay into a climate loss and damage fund to assist developing nations, having not been responsible for historical global emissions and their impact. This argument is somewhat dated now with small island developing states (SIDS) calling on China, at COP27 in Egypt in 2022, to acknowledge its contemporary responsibility as a high carbon emitter. As Chapter 7 notes, however, China is a significant climate change financial actor through its Belt and Road Initiative and South-South Cooperation Fund. Despite actively engaging in the UNFCCC and its global agreements, except for China, each of these countries is seeking recognition of their humanitarian need and support from developed countries to strengthen their climate policy capacity, to survive climate impacts and to implement policy effectively and more sustainably. Meeting their climate goals and targets is dependent on adequate international support (Chapter 5; Table 3.1).

Pacific Island Countries

As SIDS, Pacific Island countries have been at the forefront of agenda setting on the need to act as a region on climate change, to accept differentiated responsibility, to cooperate on action, to establish challenging dialogue with major emitters and to accept humanitarian regional need. Ourbak and Magnan (2018, p. 2201) observe that most recently they have succeeded in securing their special circumstances as vulnerable countries, demonstrated leadership in raising ambition to reduce GHG emissions to help secure an ambitious long-term temperature goal of limiting global warming to below 1.5 °C, and advanced the complex debate on climate loss and damage. The SIDS and the Pacific Island Community (PIC) have for decades also pressed for legitimate global emissions reduction action, calling out non-action and failed action, whilst more recently advocating for a fossil fuel non-proliferation treaty, an obligation to prevent climate harm and an acknowledgement of, and action upon, climate injustices (Chapter 8). While these nations are not themselves responsible for climate change, they are actively seeking coherent and aligned strategic measures for the region to respond to it (see Chapter 5); and engage in an active set of Pacific regional climate fora, strategies and dialogues (Chapter 2).

Assessing Capability and Differentiation

Climate Action Tracker – 'Fair Share' Analysis

In keeping with the internationally accepted norm of 'common but differentiated responsibility' between developed, lesser-developed and developing nations (Table 3.1), CAT adopts 'fair share' as a criterion by which to judge the efficacy of climate policies, actions, targets and finance. Accepting that countries have very differing 'fair share' climate efforts to make was implicit in parties signing the *Kyoto* and the *Paris* agreements, but now committing to the principle of differentiation by indeed making those efforts is, by our schema, a leadership quality. CAT has ranked Australia and Japan's efforts as 'insufficient' and New Zealand and South Korea's as 'highly insufficient' in part because they are not making a 'fair share' effort on climate change and are not meeting their responsibilities let alone leading with their actions (Chapter 1). This does not bode well for their climate leadership within their region in terms of building regional climate capacity. Making a 'fair share' contribution for these developed countries would involve steepening their NDC targets, adopting more ambitious policies, strengthening domestic emissions reductions, halting the subsidisation of fossil fuels and boosting climate finance to 'fair share', predictable levels (CAT, 2024).

Table 4.4 Capability and Differentiation – Country status, wealth and adaptive capacity

Country – status	*ND-Gain Index* (overall world rank)*	*Vulnerability (world rank)*	*Readiness (world rank)*
New Zealand – HI	70.3 (8th)	0.294 (9th)	0.701 (9th)
Australia – HI	68.9 (12th)	0.312 (19th)	0.691 (12th)
South Korea – HI	67.9 (15th)	0.370 (51st)	0.729 (6th)
Japan – HI	65.6 (19th)	0.378 (62nd)	0.690 (13th)
China – UMI	58.3 (39th)	0.387 (74th)	0.554 (36th)
Malaysia – UMI	56.9 (49th)	0.369 (47th)	0.507(54th)
Indonesia – LMI	47.6 (98th)	0.440 (103rd)	0.392 (102nd)
Thailand – UMI	52.3 (71st)	0.437 (102nd)	0.483 (62nd)
Viet Nam – LMI	47.5 (100th)	0.475 (128th)	0.426 (93rd)
Taiwan – HI	Not available	Not available	Not available

Source: DIE (2022); ND-GAIN (2021).

Notes: HI – High Income; UMI – Upper Middle Income; LMI – Lower Middle income.

*ND-GAIN summarises a country's vulnerability to climate change and other global challenges in combination with its readiness to improve resilience (see Table 5.1 for more information on EA&P countries).

Climate leaders would, we argue, acknowledge the principles of capability, or the ability to reduce emissions, to build resilience and to adapt to climate impacts, and differentiation, or the notion of differentiated efforts from developing, developing and lesser-developed nations. The *Kyoto* and the *Paris* agreements adopt this principle of differentiation, but this has not bound its signatories to acting in accordance with it. It is common sense that capacity varies between wealthier and less wealthy nations, that climate policy efforts will therefore differ and that the greatest efforts and assistance to countries with less capacity is expected from developed nations. It is relatively easy to demonstrate differing capacity, looking at differing income status, vulnerability to climate impacts and readiness to respond which we do in Table 4.4 (see also Chapter 5). It is also worth noting that leading efforts may not always relate to greater capacity, as the lesser-developed Pacific Island nations have shown by their climate action agenda setting as a region for many decades (Chapter 8).

Developed Countries - Australia, Japan, New Zealand and South Korea

The Notre Dame Global Adaptation Initiative's (ND-GAIN) Country Index (2021) details the adaptive capacity of countries worldwide (Tables 4.4 and 5.1). New Zealand, for example, demonstrated the highest adaptive capacity within the EA&P and ranked 8th overall, followed by Australia 12th; South Korea 15th; Japan 19th; China 39th; Malaysia 49th; Indonesia 98th; Thailand 71st; and Viet Nam 100th. There is no data available for Taiwan. While higher ranking developed countries provide some financial support for climate mitigation and adaptation, none of it is sufficient to respond to need (Chapter 5). Historical responsibility and acknowledgment of differing capacity is not recognised in NDCs, but neither are climate finance, fairness, equity, loss, damage and human rights, nor technology transfer mentioned. In practice, developed countries have provided at least some, if inadequate, financial support for capacity building and technology transfer. Japan contributed US$677 million to the Global Environment Facility (GEF) for building the capacity of lesser-developed and developing countries and assisting technology transfer, followed by Australia (US$87 million), New Zealand (US$11 million) and South Korea (US$10 million) (DIE, 2022). There is no clear way of assessing leading actions other than to note that the dollar level of developed country support for climate action in EA&P is considered grossly inadequate (Lim et al., 2024).

Malaysia, Taiwan, China, Indonesia, Thailand and Viet Nam

There is little data available from Malaysia, Taiwan, China, Thailand and Viet Nam. Fairness, equity, loss and damage and human rights are not in their NDCs. Capacity building and technology transfer are not mentioned either. There is no data available for Taiwan (Donor Tracker, 2023).

Pacific Island Countries

Relatively speaking, Pacific Island countries have little capacity to respond to, or build resilience against, the impacts, and accelerating impacts of climate change which are an existential threat to their nations. Table 1.2 details their relative poverty, whilst Table 5.1 details their extreme vulnerability and exposure to climate risk, as well as their relative lack of climate change preparedness, and their lack of adaptive capacity and governance capacity. These are not countries with 'structural leadership' as depicted by the climate leadership literature because they have tiny domestic economies and are not economic powerhouses with vast financial leverage (Chapter 8). Pacific Island countries certainly acknowledge their difficulties and lack of capacity and use these not to excuse a lack of their own climate action, but to leverage enhanced action from developed and major emitting countries as well as in international negotiations. They expect to be the net recipients of climate finance, including for climate loss and damage, capacity building support, and technology transfer from wealthier countries, particularly those with historical responsibility for driving climate change.

Conclusions

In this chapter, we assessed the efforts of select developed, lesser-developed and developing EA&P countries against our regional climate leadership criteria of credibility, effectiveness, responsibility, coherence, capability and differentiation. We acknowledge that CAT and CCPI have already identified the poor performance of these countries, notably from the developed countries, in terms of meeting the aims of the *Paris Agreement*. We discuss our findings in detail in Chapter 8; however, developed countries are clearly undermining their *Paris* aligned policy goals by not cutting emissions. Any collaborative efforts they may be making on tackling climate change within the EA&P region, which we are looking for as a measure of regional leadership, is undercut by this stark reality. Instead, there is reliance in developed countries upon offsetting, ineffective mitigation mechanisms, and on fossil fuels, with no overt efforts to transition to low carbon economies sufficient to achieve their *Paris* goals. The lesser-developed countries we considered exhibit most of these same shortcomings, and poor CAT and CCPI rankings. However, countries from Malaysia to the Pacific are also voluntarily pursuing their *Paris* goals, legislating action, establishing carbon pricing and policy exchanges, engaging on climate action across the region and adopting energy transition roadmaps.

There may be scope within the NDC process for declaring nation state actions that align with, and nation state actions that undermine, the goals of the *Paris Agreement*, and scope for declaring which actions by developed countries demonstrate a bona fide acceptance of common but differentiated responsibility. We have focused, for example, on the responsibility, which developed countries are not exhibiting, to contribute sufficient financial support to lesser-developed and developing countries in the EA&P region in their efforts to reduce emissions, build capacity and adapt to climate impacts. The task of building capacity and achieving resilience is found to be a challenging one given the EA&P's extreme vulnerabilities, as Chapter 5 finds, with the burden falling on the less capable nations with no historical responsibility for climate change. We did note significant levels of engagement between EA&P countries and China, which we see in Chapter 7 is the only economically significant country within the region to be overtly assuming the mantle of domestic and regional climate leader. However, we discuss our regional climate leadership findings in full in Chapter 8.

5 Preparedness, Adaptation and Policy Resilience in EA&P

Introduction, Climate Impacts, Vulnerability and Risk

In this chapter, we consider the daunting scale, and urgency, of the adaptation and resilience building challenges across East Asia and Pacific region (EA&P) and argue that leading cross-sector polycentric global, domestic, regional and local actions and policy solutions are needed. As we saw in Chapter 4, domestic and regional climate leadership efforts should be a part of responding to these challenges. Whilst the EA&P is currently, although not historically, the world's highest regional contributor to increased emissions, it is also already greatly impacted by climate change and is set to experience increased natural disasters from global warming but with little capacity to respond at the scale required (UNESCAP, 2022). As we saw in Chapter 4, there is already a high level of vulnerability and a low level of resilience in the region, which has been exacerbated by the COVID-19 pandemic, coupled with a deficit in governance and financial capacity, and in technological capacity in terms of emissions reduction (see, for example, Table 5.1). Furthermore, increased natural and biological hazards will exacerbate the pre-existing multidimensional risks for the people and economies of the region and act as a threat multiplier to those already in poverty (WMO, 2023).

Here we find that climate change and risk preparedness in general, let alone leading actions, is a complex task on several fronts in the face of the immense and intensifying scale of the vulnerability in EA&P. Firstly, there is the sheer scale of the risk and the varying abilities of EA&P countries to respond, as Chapter 4 confirms. Secondly, whilst the building of resilience is now a doctrinal response, the term resilience is a contested one. This is confusing and frustrates its conversion into an efficacious tool, let alone a leading tool for policy action (De Bruijne et al., 2010, p. 28). Thirdly, the financing of the response to climate impacts and risks by developed countries is manifestly inadequate. The Asia-Pacific region requires US$362 billion per annum (UNESCAP, 2023c, p. 51), but the *Paris Agreement's* global goal is to mobilise US$100 billion per annum, and only a fraction of this is being achieved. These complexities need to be acknowledged and woven into the cascading sets of frameworks for action which provide the context for action that should

DOI: 10.4324/9781003170389-5

Table 5.1 Select EA&P countries – vulnerability, adaptive capacity, readiness, governance and risk

Country	*Income group*	*Vulnerability*	*Readiness*	*Adaptive capacity*	*Governance*	*Climate risk**
New Zealand	Upper	0.294 – 9th	0.701 – 9th	0.204 – 4th	0.869 – 3rd	66.33 – 66th
Australia	Upper	0.312 – 19th	0.691 – 12th	0.297 – 25th	0.812 – 12th	28.00 – 19th
Malaysia	Upper Middle	0.369 – 49th	0.507 – 54th	0.411 – 45th	0.581 – 62nd	87.33 – 99th
Rep. of Korea	Upper	0.370 – 51st	0.729 – 6th	0.328 – 28th	0.684 – 30th	64.00 – 60th
Singapore	Upper	0.375 – 59th	0.805 – 1st	0.284 – 24th	0.893 – 1st	118 – 130th
Japan	Upper	0.378 – 62nd	0.690 – 13th	0.276 – 21st	0.782 – 16th	14.5 – 4th
China	Upper Middle	0.387 – 74th	0.554 – 36th	0.439 – 59th	0.471 – 91st	42.83 – 32nd
Fiji	Lower Middle	0.445 – 114th	0.478 – 67th	0.598 – 112th	0.573 – 64th	73.00 – 75th
Laos	Lower Middle	0.460 – 117th	0.336 – 136th	0.670 – 139th	0.398 – 128th	55.17 – 45th
Philippines	Lower Middle	0.463 – 121st	0.337 – 135th	0.511 – 100th	0.405 – 125th	26.67 – 17th
Viet Nam	Lower Middle	0.475 – 128th	0.426 – 93rd	0.561 – 102nd	0.454 – 101st	50.17 – 38th
Palau	Upper Middle	0.485 – 131st	0.427 – 92nd	No Data	0.520 – 76th	No Data
Cambodia	Low	0.486 – 132nd	0.288 – 159th	0.729 – 153rd	0.353 – 145th	75.83 – 84th
Myanmar	Low	0.504 – 140th	0.257 – 178th	0.646 – 129th	0.227 – 177th	31.33 – 21st
Samoa	Lower Middle	0.507 – 143rd	0.439 – 84th	0.601 – 113rd	0.621 – 47th	No Data
Papua New Guinea	Low	0.546 – 160th	0.283 – 164th	0.784 – 167th	0.356 – 142nd	58.67 – 51st
Vanuatu	Low	0.556 – 164th	0.384 – 105th	0.627 – 121rd	0.552 – 71st	73.00 – 75th
Nauru	Lower Middle	0.581 – 171st	0.493 – 59th	No Data	0.493 – 81st	No Data
Solomon Is	Low	0.599 – 176th	0.396 – 100th	0.680 – 143rd	0.784 – 167th	118 – 130th
Tonga	Lower Middle	0.605 – 179th	0.426 – 93rd	0.572 – 106th	0.534 – 73rd	118 – 130th
Micronesia	Low	0.616 – 181st	0.359 – 118th	No Data	0.569 – 65th	118 – 130th

Sources: Global Adaptation Index (ND-GAIN, 2022); Global Climate Risk Index (Eckstein et al., 2021).
Note: *The countries ranked the highest are the most impacted by extreme climate events in 2019.

be driven by developed countries. We also argue that the resilience of climate policy design is a neglected aspect of climate change preparedness but crucial to meeting its challenges and effecting much needed change.

Climate Impacts, Vulnerability and Risk

In Chapter 1, we detail the extreme impacts of climate change in the EA&P region, a region that is already the most disaster stricken in the world, with climate change impacts outstripping adaptation efforts whilst increasing natural disasters and involuntary population migration. The toll already includes warming winds, increased heatwaves, droughts, cyclones, floods, typhoons, glacier melt, sea level rise and oceanic warming, biodiversity loss, human diseases and deaths, as well as increased food and energy insecurity, industrial systems risk and economic vulnerability (IPCC 2022b; WMO, 2023). The situation is even more acute because warming in the region, which is already disproportionately hazardous – with the region therefore disproportionately vulnerable in global terms – has exceeded the global mean, putting at risk regional environments, economies, and societies, and is set to continue. The United Nations Economic and Social Commission for Asia and the Pacific (UNESCAP) warns that the impacts, crises and disasters that are already being experienced in the region are therefore fundamental, continuous, multiple, overlapping, outstripping capacity and accelerating (UNESCAP, 2023c).

Unsurprisingly, the University of Notre Dame's Global Adaptation Index's (ND-GAIN) leader board of climate prepared countries shows the lower income EA&P countries, such as the Pacific nations of Micronesia, Tonga and the Solomon Islands, as the most vulnerable to climate change, the least ready to meet its challenges, with the lowest adaptive capacities, and generally with less governance capacity (Table 5.1; ND-GAIN, 2022). Upper, to upper middle, income countries such as New Zealand, Australia, Malaysia, Rep. of Korea, Singapore and Japan are vulnerable to climate risks, but with higher levels of readiness, adaptive capacity and effective governance. Despite its vast size and population, economic promise, and the growth of its middle class, China has a more heightened vulnerability to climate change, a lower adaptive capacity and a much less effectual system of governance. The more 'climate capable' countries in the region have better scope for climate leadership, with New Zealand and Singapore leading in climate governance and in the building of readiness and adaptive capacity. However, even the most climate-prepared countries are losing ground in terms of readiness, with some of the highest-ranked, and higher income, countries, such as Japan, seeing a sharp increase in their vulnerability to climate change (Beasley, 2022).

The Global Climate Risk Index (CRI) finds that between 2000 and 2019, 'over 475,000 people lost their lives worldwide and losses of US$2.56 trillion were incurred as a direct result of more than 11,000 extreme weather events'. Slow onset processes will add significantly to this (Eckstein et al., 2021, p. 6). The index (see Table 5.1) draws upon past data of extreme events and associated socio-economic data, to estimate the direct impacts (losses and fatalities) rather than the indirect impacts of a warming climate, upon agricultural capacity, for example. Myanmar and the Philippines as amongst the EA&P countries most impacted by extreme weather events from 2000 to 2019, and Japan amongst those most affected in 2019. That year Japan was hit by two typhoons, one the most powerful in 60 years, as well as a series of extreme heatwaves, with May temperatures in Hokkaido the highest recorded. It also suffered 157 deaths, and losses and damages totalling US$25 billion. The capacity to recover from and adapt to further extreme events is challenging for countries such as Myanmar and the Philippines that are now the most affected year on year (Eckstein et al., 2021, pp. 6, 10, 15). In 2022, the global costs of extreme events were estimated at US$160 billion to $340 billion per annum by 2030, and US$315 billion to US$565 billion per annum by 2050, with residual risks and unavoidable loss and damage likely to double these amounts (UNEP, 2022).

What Is Climate and Disaster Resilience?

The impacts of climate change are an added burden on the already disaster-prone EA&P, so it is unsurprising that the term 'resilience' has been acknowledged as crucial as both an abstract aspirational concept and as an applied policy outcome. By the 1990s, when global climate governance fostered by the United Nations was just beginning to take shape, the term resilience was already widespread as a key component, indeed a key aim, of disaster risk reduction (DRR). In 2002, for example, the *United Nations International Strategy for Disaster Reduction* (UNISDR) defined resilience as 'the capacity of a system, community or society to resist or change in order that it may obtain an acceptable level of functioning and structure'. Subsequently, in 2005, the United Nations sponsored *Hyogo Framework for Action 2005–2015* explicitly recognised the need for a comprehensive global approach to disaster risk reduction that is part of a sustainable development approach and, crucially, that is integrated across all sectors and disciplines and fundamental to climate change adaptation policy (CCAP). One of the primary objectives of the Hyogo framework, and its successor, the *Sendai Framework for Disaster Risk Reduction 2015–2030*, has been to build resilience at national, regional and local community levels (Kelman, 2015).

Building resilience is a challenging task, however, not least for its fractured definitions with iterations of the term contested in the fields of its application as diverse as climate science, disaster management, engineering, ecology, social science, psychology and public policy. De Bruijne et al. nevertheless find that the term across these various fields means essentially what one would expect – that is, deriving from the Latin *resilio* or to 'jump back'. Resilience is the flip side of vulnerability, in other words it is about resisting disorder, and 'emphasises the ability of systems or persons to cope with hazards and provides insights on what makes a system more or less vulnerable'. Resisting adversity and dealing with uncertainty and change are key components of resilience (De Bruijne et al., 2010, pp. 13–14). Despite the fractured, multidisciplinary application of the term resilience, there is a commonality to its understanding as resisting harm and returning to equilibrium in the physical sciences, in engineering, and in the ecological sciences, where ecosystemic resilience is emphasised. Key to more recent notions of climate and disaster resilience are socio-ecological resilience, economic resilience and social or community resilience, in terms of building resilience in dealing with disturbance and stress, and responding to, learning from, and adapting to various types of change (Adger, 2000; Gunderson & Holling, 2002; Gallopin, 2006; Reid & Botterill, 2013; IPCC, 2014).

A further challenge in terms of building resilience in EA&P is the difficulty not only of applying varying definitions of the term but also of adopting an integrated approach in the design of climate change mitigation and adaptation and disaster risk reduction frameworks. Kelman (2015, pp. 125–6) finds, for example, that the *Sendai Framework for Disaster Risk Reduction* fails to frame climate change as a driver of disaster and an enhancer of vulnerability and that, at least until 2030, climate change will be treated entirely separately from disaster risk reduction. There had been initial international recognition in the Bali Action Plan, by leading developed nations, as parties to the *United Nations Framework Convention on Climate Change*, of the need for disaster risk reduction strategies to be considered tools for climate adaptation (UNFCCC, 2007). However, the United Nations eventually identified a more sophisticated aspiration for integration whereby the overlapping goals of climate change adaptation, disaster risk reduction and sustainable development agendas would be to reduce vulnerability and to enhance resilience (UNCCS, 2017).

The current UNESCAP (2022) approach to responding to the 'riskscape' of 1.5 °C is now to integrate *disaster risk reduction*, *climate change adaptation* and *resilience building* mechanisms into regional and sub-regional pathways that are specific to local conditions.

The practical application of well-integrated adaptation and resilience frameworks, planning and policies affords clarity in terms of how 'resilience' is to be defined and of any common or differential treatment of action on climate change and disaster risk reduction. The *Climate Smart Disaster Risk Management* approach, for example, rests upon various integrated pillars of action that acknowledge that multifaceted risks and multiple, often simultaneous, shocks and stressors are part of lived community reality where impacts are felt. Policies, plans and funding that ignore this reality, and indeed that are not operationalised according to the specific local context at hand, will fail to reduce vulnerability and poverty and to build resilience (Bahardur et al., 2010). The Asian Development Bank also has an integrated approach to climate and disaster resilience building. It defines the goal of resilience as multifaceted – with eco-based, physical, financial, social and institutional elements, and integrated with actions to mitigate climate change and to pursue sustainable development (ADB, 2019). Its efforts to support infrastructure development, disaster risk management and sustainable urban development are critical in the region. In terms of leading, or best practice approaches, UNESCAP (2022) also advocates integrating adaptation and risk reduction for differing risks in differing contexts.

Although the impacts of climate change and the occurrence of natural disasters are intensifying in the Asia Pacific, and despite failed efforts to raise sufficient finance to mitigate the problem, there is at least an appreciation now of the crucial interrelationship between climate and disaster.

Financing Resilience in EA&P

The financing of mitigation, adaptation and resilience building is therefore crucial. The insurance industry is aware that the accelerating impact of climate change is exposing the under-insured and financially vulnerable Asia Pacific to the unaffordable costs of increasingly volatile weather with record-setting temperatures, intensifying rainfall, flooding, droughts, fires, cyclones and storms. In 2022, natural disasters and catastrophes in the Asia Pacific were estimated to cost US$80 billion in insurance costs, with flooding dominating at 60% of the loss total, with one third experienced in Australia (predominantly from flooding), and record rainfalls and temperatures across Asia (Cheeseman, 2023). The Swiss Re insurance group has ranked 48 economies according to how severely their GDP is likely to be impacted by climate risk factors by 2050, without resilience building to mitigate these risks, and has found that climate change will likely cost Asia more than a quarter (26.5%) of its GDP by 2050 (noting, however, that this is an insurance company estimate). Unsurprisingly, the developing, lower income Asian economies such as Thailand, India, the Philippines, Malaysia and Indonesia, with the lowest adaptive capacity and the greatest vulnerability to extreme weather, will experience the greatest risk and impact on their GDP (Higginbotham, 2021; Table 5.1).

Climate finance is therefore crucial for EA&P, and is urgently needed, to accelerate mitigation and adaptation, and to build capacity and resilience in dealing with climate impacts. But it is not adequately forthcoming from developed countries. The United Nations defines climate finance as 'local, national or transnational financing—drawn (in polycentric fashion) from public, private and alternative sources of financing—that seeks to support mitigation and adaptation actions that will address climate change'. The acknowledgement of the need for financial assistance for developing and lesser-developed countries from developed countries has been a feature of international climate change conventions, protocols and agreements for over three decades. The *United Nations Framework Convention on Climate Change*, the *Kyoto Protocol* and the *Paris Agreement* all make the call for developed countries to provide climate finance to lesser-developed and developing countries which are 'less endowed and more vulnerable'. To facilitate the provision of such finance, the UNFCCC established the Global Environment Facility (GEF) in 1994 and the Green Climate Fund (GCF) in 2010, as well as a Special Climate Change Fund (SCCF) and Least Developed Countries Fund (LDCF).

The *Paris Agreement* Parties confirmed a global goal of mobilising US$100 billion in climate finance per year by 2020, with a steeper goal to be set by 2025 (UN, 2023a). However, achievements are well short despite leading contributions from Japan, South Korea and Australia. Since 2015, the OECD has tracked the finance from developed countries, aid agencies, private agencies, development banks and climate-related supported export credits. It finds that in 2020 only US$83.3 billion was mobilised jointly for climate action in lesser-developed and developing countries, mainly for mitigation, but with growing investment in adaptation, primarily as loans mainly targeted at Asia and middle-income countries (OECD, 2022a). But NGOs such as Oxfam, a movement against inequality and injustice, allege that developed countries are significantly inflating the value of their climate finance, that the finance has less climate-focus than stated and that loans, rather than grants, are adding to the debt crises of lesser-developed and developing countries. And that 'while developed countries claim their climate finance provided and mobilised reached $83.3bn in 2020 ($13.1bn of which was mobilised private finance), Oxfam estimates the actual value of climate assistance provided to lesser-developed and developing countries to have been one-third of that' (Oxfam, 2022).

The IPCC (2022b) provides a synthesis of the peer-reviewed literature on climate finance, which confirms its shortcomings, although it notes the literature is not up to date nor comprehensive and suffers from underdeveloped and conflicting methodologies (Ahmed et al., 2019). Adaptation cost estimates have improved; however, projected GDP costs to Asian economies are a fraction of those proposed by the insurance industries, whilst regional level cost analysis largely derives from analysis by institutions such as the Asian Development Bank (ADB, 2023). The literature confirms that funds established such as the GCF, SCCF, and notably the LDCF, are channelling significant climate adaptation funding into EA&P but that this is often for marginal, not transformational, projects. Despite the emergence of a range of new developing-country finance institutions with mandates to provide adaptation and resilience finance, the world is not achieving the US$100 billion per year climate finance goal that was endorsed by the *Paris Agreement* (IPCC, 2022b). UNESCAP (2023c, p. 51) estimates[1] that lesser-developed and developing countries in the Asia Pacific are receiving only one tenth of the US$362 billion per annum that the region requires to meet its *Paris Agreement*-based mitigation and adaptation goals. To achieve its 2030 goals, the region's climate finance needs to be boosted by c.590% (CPI, 2021).

The financing of climate resilience in EA&P is a challenge, therefore, on multiple fronts, beyond meeting specific domestic financial investment challenges (see Table 5.2), the principal challenge being to reverse under-investment from the region's developed countries. Regional finance is urgently needed and needed at a scale that dwarfs the global climate finance goals set under the *Paris Agreement*. Furthermore, where climate funding is available for the region, it is

Table 5.2 Climate finance in select EA&P and developed countries

Country	*GDP*	*Climate finance*	*CAT assessment of climate finance*
United States*	$73,600	Critically insufficient	Increased commitment but contributions very low compared to 'fair share'; must end fossil fuel funding
European Union*	$45,830	Insufficient	Contributions low compared to 'fair share'; need to phase out the financing of international fossil fuels
Singapore	$93,400	Not CAT assessed	No contributions to the UN Green Climate Fund
Australia*	$59,500	Critically insufficient	No contributions to the UN Green Climate Fund; low levels of funds/ bilateral finance; funds fossil fuels
New Zealand	$48,800	Highly insufficient	Contributions very low compared to 'fair share'; need to prevent funding of fossil fuels abroad
South Korea*	$50,600	Not CAT assessed	Contributed US$600 million to UN Green Climate Fund[1]
Japan*	$46,300	Highly insufficient	Contributions very low compared to 'fair share'; need to prevent funding of fossil fuels abroad
Thailand	$21,100	Not applicable [CAT]	Thailand would expect to be a recipient of funds
China*	$22,100	Not CAT assessed	No contributions to the UN Green Climate Fund. Paid 10% so far of the $3.1 billion it pledged to China's South-South Climate Cooperation Belt & Road Fund since 2015[2]
Indonesia*	$14,100	Not applicable [CAT]	Indonesia would expect to be a recipient of funds

Sources: GPD per capita calculations by CIA (2024); Climate finance assessment by CAT (2024); other sources – [1]Haye-ha (2023); [2]E3G (2023) – E3G is an independent climate change think tank - The Third Generation Environmentalism Ltd.
*denotes G20 countries.

a fraction of what is needed, and predominantly in the form of loans that can exacerbate the debt crises of lesser-developed and developing countries, without necessarily having any transformational impact. Capacity building is urgently required, UNESCAP finds, to ensure data availability and comparability, to establish a coherent policy environment, to ensure the bankability of high-risk climate projects, to clearly align financing with national climate goals, to assist with regulatory compliance and to support the introduction of innovative climate financial instruments such as green bonds and carbon pricing (UNESCAP, 2023c, pp. 49–59). After decades of talk, there was a breakthrough at the meeting of the parties to the UNFCCC, COP27 in Cairo, with an agreement to provide 'loss and damage' funding for vulnerable countries hit hard by climate disasters, with a transitional committee to meet to consider recommendations on operationalising the new arrangements (UN, 2022a).[2]

Assessing Regional Preparedness

Integrated approaches to climate and risk management do not necessarily align to achieve best practice, indeed leading regional preparedness in EA&P, including with international frameworks. Given the 'cascading risks in the Asia Pacific region', UNESCAP has assessed such risks in its mid-term review of the *Sendai Framework for Disaster Risk Reduction 2015–2030* and its implementation. This framework has seen a reorientation from disaster response to preparedness and is intended to be adopted across 'policy planning processes and actions within social, economic, and environmental systems at the regional and domestic level' within the Asia Pacific. The UNESCAP Committee on Disaster Risk Reduction recognises that a systematic approach is required across the region to build capacity, and to scale up integrated regional and sub-regional climate-related disaster strategies. The committee recognises that this will require a country level commitment to operationalising global climate and risk frameworks in regional and sub-regional plans in terms of policy coherence, integrated climate-change and risk-based adaptation, and resilient infrastructure system planning. However, the *Sendai* mid-term review has not found this commitment in many Asia-Pacific countries (UNESCAP, 2023d).

In Chapter 2 we saw that the goal of building resilience in ASEAN countries is prominent in ASEAN's climate change vision, statements and action plan, and its cooperative work and bilateral statements with partner countries in the Asia Pacific. Prominent, facilitative regional forums and groupings (Table 2.1) include ASEAN's Working Group on Climate Change, ASEAN's Climate Resilience Network and the ASEAN Regional Forum's Intersessional Meetings on Disaster Relief (ISM on DR). In the Pacific, the prominent forums and communities include the Pacific Islands Forum (PIF), the Pacific Islands Forum Secretary General's Young Climate Leaders Alliance, the UNSG's Youth Group on Climate Action, the High-Level Advisory Group on Climate Action and the broader SPC Pacific Community which includes Australia, France, New Zealand, the United States and the United Kingdom. The climate change language in the Pacific Island's vision, statements and action plan prioritises the need for building climate change resilience, with a Pacific Resilience Partnership established by PIF to ensure the implementation of climate and disaster risk management. The processes are in place, but effective financing for implementation is not assured.

Developed countries are not leading in terms of providing funding for regional climate and disaster risk management. Their support is patchy and inadequate compared to the funds that are required. Japan's funding efforts are 'highly insufficient' and include the funding of fossil fuels (CAT, 2024), but it has driven preparedness in ASEAN countries in part through the disaster-based research efforts of the Japan-ASEAN Integration Fund (JAIF). It has supported projects such as regional risk financing and insurance, the establishment of a Disaster Emergency Logistic System for ASEAN (DELSA), and disaster relief exercises. JAIF also supports cooperative efforts to reduce risk and to promote the adoption of early warning systems, as well as adaptation and capacity building through the ASEAN-Japan Disaster Reduction Plan. And despite its CAT rating of 'critically insufficient' climate funding, for withdrawing from the GCF and for funding fossil fuels, Australia is marginally assisting resilience efforts in the Pacific, for example, with its AUD$700 million pledged in resilience project assistance 2020–25, although its resilience building funding is not wholly climate change specific. The current Australian government will strengthen support for regional disaster preparedness and response, for building climate resilient infrastructure and for increased access to renewable energy (DFAT, 2023a, 2023b).

However, for climate finance to be effective in building preparedness, it must be systematically integrated and coherent in a policy sense, and in sub-regional, national and local contexts with the cascading sets of frameworks for building climate and risk resilience. To this end, the *Sendai Framework* calls for the development of regional action plans and strategies. In EA&P the relevant regional framework is the *Asia-Pacific Action Plan 2021–2024 for the implementation of the Sendai Framework for Disaster Risk Reduction 2015–2030* (UNDRR, 2021). This framework aims 'to accelerate Asia Pacific's transformation to risk-informed development, by treating disaster risk reduction as a cross-cutting theme and by increasing investment in prevention, risk reduction, climate change adaptation and anticipatory approaches to enhance resilience'. In the Pacific, the relevant implementing framework is the *Framework for Resilient Development in the Pacific: An Integrated Approach to Address Climate Change and Disaster Risk Management 2017-2030* (FRDP) which has the goals of strengthened integration adaption and risk reduction, low carbon development and strengthened disaster preparedness, response and recovery.

Post-2015, however, of the 195 signatories to the *Sendai Framework*, only one country is at the stage of having its implementation validated, with 131 countries not having even begun the process. A further shortcoming is the lack of climate specific or context specific indicators in the framework's monitoring process. Noting, however, that by 2021, 121 countries had reported having implemented national disaster risk reduction strategies (UN, 2022b).

Regionally and sub-regionally specific climate and risk preparedness planning must also align with the United Nations Sustainable Development Goals (SDGs) to minimise climate impacts and vulnerability to disaster. UNESCAP has found that whilst there has been some progress in South-East Asia and the Pacific towards some of these goals, they are nowhere near to being achieved, and there has in fact been negative progress on climate change action, and the building of resilience and adaptive capacity. Those are nevertheless issues that the *Asia-Pacific Action Plan* for implementing the *Sendai Framework for Disaster Risk Reduction* 2015–2030 prioritises – for better understanding, for strengthened more systematic governance, and for better regional complementarity 'between climate change policies and actions, especially National Adaptation Plans, disaster risk reduction strategies and national COVID-19 recovery efforts'. In the post-pandemic context, decisive action to implement policy framed by sustainability aspirations and regional co-operation will aid recovery from the crisis across affected sectors with immediate positive effect (UNESCAP, 2020; UNDRR, 2021, 2024).

And finally, also of relevance is the *Asia-Pacific Action Plan*'s identification of the need for the adoption of best practice climate and disaster policy design (UNESCAP, 2022).

Developing Climate and Disaster Policy Resilience

Resilience is, by definition, a complex, contested term with a wide variety of applications in vastly differing systems, each of which, arguably, is relevant to building a robust capacity to better withstand, and to bounce back more readily from, climate change and disaster impacts. Typically, resilience is thought of as systems with differing 'resilience profiles' – 'physical, biological, psychological, social and cultural', but not at all in terms of political systems nor the policies that promote or inhibit resilience building (Birkland, 2010, p. 106). Political decision-making will impact the types of mitigating projects, actions, processes and policies that are adopted which, Birkland finds in terms of his analysis, may improve or indeed undermine resilience. There are many well-documented shortcomings in the building of resilience, he finds, where efforts undermine each other, for example, where community safety efforts result in inappropriate actions with longer term adverse consequences. In a macro sense, there is also the well-known problem in his example of the United States, where the disaster policy emphasis is upon relief and insurance pay-outs, rather than upon preparedness and mitigation (pp. 106–110).

If climate change and disaster risk resilience are to be achieved, both domestically and within the region, then in addition to the features of best practice, well integrated, preparedness at multiple levels, the design of policy and the policy process itself is significant. Birkland (2010, p. 109) suggests that differing phases of the disaster cycle – preparedness, mitigation, response, and recovery – need to be thought of 'as aspects of a system that promotes or retards resilience'. The problems with resilience are multifaceted, but they are well documented and include these types of issues: (i) the need for explicitly strong leadership from government; (ii) the need for top-down bottom-up integrated policies and action; (iii) the need to consider the appropriate scale and applicability of policies and action; (iv) the need for community debate and collaborative decision-making in determining policies and action; (v) the need for future orientation so that current policies and action anticipate future consequences; (vi) and the need for explicitly designed policy process (Table 5.3) from formulation to implementation and evaluation. The policy process itself must

Table 5.3 Achieving resilience through policy design

Resilience defined for monitoring?	Policy instruments identified?
Vulnerability baselines set?	Outcomes, targets, goals set & monitored?
Actors identified and engaged?	Implementation & evaluation undertaken?
Deliberation/involvement processes adopted?	Learning identified for policy transfer?
Barriers identified? [data; gaps; obstacles?]	Leading resilience building actions achieved?

Source: Nakamura and Crowley (2016).

be resilient and not neglectful of baselines, of monitoring, of engagement, of likely pitfalls and obstacles, of targeting, of implementation, of evaluation and of learning, as aspects, although not necessarily as sequential aspects, of the process (Nakamura & Crowley, 2016).

ASEAN's *Asia-Pacific Action Plan 2021–2024 for the implementation of the Sendai Framework for Disaster Risk Reduction 2015–2030* advocates the building of regional and national level resilience through robust policy design. It aims to adopt and promote coherent planning and implementation of national and local disaster risk reduction and climate change adaptation strategies, including enhancing the linkages between national development plans, NDCs, disaster risk reduction strategies, National Adaptation Plans and national COVID-19 recovery efforts, while optimising effective disaster risk reduction measures across all policy sectors with an emphasis on proactive pre-disaster investments. It advocates the formulation of monitoring, evaluation and accountability frameworks to monitor implementation and to evaluate the effectiveness of the national and local disaster risk reduction and climate change adaptation strategies. Climate change is acknowledged as an exacerbating threat, with intensifying impacts, and importantly it is acknowledged that EA&P is not on track to achieve any of its sustainable development goals. The scaled-up implementation of the *Asia-Pacific Action Plan* has the capacity to reverse this trend (UNDRR, 2024). It is mirrored by equally significant and detailed resilience planning by the Pacific community.

The *Framework for Resilient Development in the Pacific* integrates for the first time the region's previous approaches to climate change adaptation and disaster risk management, with learnings from these approaches, and from extensive engagement. It is a coordinated regional approach to providing high level strategic guidance to member countries and stakeholders on a set of non-exhaustive priority actions, with the caveat that regional, national and community level implementation will vary according to vulnerability, need, capacity and context. There is guidance for climate and disaster integration and its mainstreaming into development planning, including policy making, planning, financing, programming and implementation, in efforts to build resilience in the Pacific. Capacity will be built by strengthening and developing polycentric partnerships across countries and territories, between stakeholders and the community, and by sharing lessons without compromising sovereignty. It offers coordinated support to the region in a context that aligns with the need in the region, its countries and communities, as well as with global and regional agreements, agreements and agendas. It also provides for monitoring national actions against a framework aimed at achieving Pacific resilience, and for measuring progress against the key pillars of Pacific resilience, i.e. to *Integrate*, *Inform*, *Include* and *Sustain*. A monitoring, evaluation and reporting framework will be developed with support from regional organisations and development partners, and a mid-term review of the framework is due in 2024 (Pacific Community, 2016).

Conclusions

The vulnerability of EA&P to climate change and the damage that is being experienced are reasons for urgent attention to, and investment in, regional preparedness, adaptation and resilience. The scale of the threat is existential for many EA&P communities. But, as we saw in Chapter 1, Chapter 4 and in Table 5.1, there are differing capacities to respond, and to finance a response, in developed, lesser-developed and developing countries. Hence, our emphasis upon polycentric state and non-state action to build capacity but also upon regional and domestic leadership. Our regional climate leadership criteria (Crowley & Nakamura, 2017) have thus been extended to include 'capacity and differentiation', to acknowledge – country capacity, ability to support climate finance, cognisance of the needs of the region, support for lesser-developed and developing countries, and support for technology transfer (Chapter 3). The obstacles to climate leading resilience building are many. The task must be defined, financed and integrated between sites of action, with community and stakeholder engagement in terms of focus, and well-designed policies that are monitored, implemented, evaluated and learnt from.

In assessing regional preparedness, we have found that the climate 'riskscape' (UNESCAP, 2022) in the Asia Pacific is appreciated and that solutions are well articulated. The challenge is well captured not only by the WMO and IPCC but also by NGOs and the banking, finance and insurance industries, and at the regional level in ASEAN's, PIF's and SPC's climate action plans. It is articulated in policy frameworks more broadly, with an emphasis now on integration with global sustainability goals and the climate ambitions of the *Paris Agreement* and the disaster risk reduction focus of the *Sendai Framework*. Preparedness, adaptation and resilience building will, however, be undermined by not rapidly advancing low carbon development, which is a centrepiece aim of the *Framework for Resilient Development in the Pacific*. We saw in Chapter 1 that decarbonisation in the Asia Pacific needs to be accelerated *by a factor of at least ten* to avert locking in dangerous climate change (World Bank, 2021).

This will require a rapid transition from fossil fuels and urgent decarbonisation efforts by EA&P states such as China, Australia and Indonesia which are economically reliant upon coal consumption, production and export, to which we now turn.

Notes

1 As a percentage of sub-regional GDP, the highest adaptation cost is estimated for the Pacific SIDS.
2 See 'Loss and damage' help for countries worst hit by climate crisis, https://www.theguardian.com/environment/2023/nov/05/countries-agree-key-measures-to-fund-most-vulnerable-to-climate-breakdown?

6 Regional Action and Carbon Pricing in the EA&P

Introduction

In Chapter 2, we saw that it is the actions of China, India, Indonesia, Viet Nam and Japan that could push global carbon neutrality beyond reach with these countries responsible for 80% of the world's new coal plants and 75% of existing coal capacity. We saw that coal consumption trends need to reverse, notably in China and India, the region's biggest emitters, and production must be reined in by China, Indonesia, India and Australia. But we also saw that the capacity for transitioning to net zero is robust, given that wealth in the Asia Pacific has quadrupled between 2000 and 2020, driven largely by China, but also Australia and Japan, and that climate action is economically stimulatory. Except for Viet Nam, these are all G20 countries with the potential for climate leadership, committed to net zero emissions by 2050, and aware that carbon pricing, either as a tax or an emissions trading system (ETS), is mooted as a cost-effective way of delivering deep emission cuts. The IMF explains that other regulatory or sectoral specific policies or actions may be more politically feasible, but would be less effective, more expensive to implement and slower to have an impact, if indeed they do, on reversing global warming (IMF, 2021a).

However, none of these countries yet has a record of cutting emissions using carbon pricing, and they all rank poorly on climate action and emissions reduction, and nearly all on emissions growth post 1990 (Table 1.2). Climate Action Tracker (CAT) and the Climate Change Performance Index (CCPI) also find that they have no viable plans for exiting fossil fuels (Chapter 4). Neither are the wealthier of these countries helping the region sufficiently finance a transition to net zero (Chapter 5). The G7, Denmark and Norway are pursuing *Just Energy Transition Partnerships* with Indonesia (US$20 billion) and Viet Nam (US$15.5 billion), although agreeing on exiting coal is slowing progress (Nangoy & Vu, 2023). In this chapter, we consider the role of carbon pricing in East Asia and Pacific (EA&P), domestically and as a regional initiative, for transitioning towards a lower carbon economy. We review carbon pricing as a mitigation measure, the state of carbon pricing in EA&P, and the design features for effective action that might be expected from climate leaders. Carbon pricing is certainly not a policy panacea, however, nor a standalone

DOI: 10.4324/9781003170389-6

measure, but a complex tool with merits and limitations, depending upon its coverage, its impacts and its efficacy.

Carbon Pricing as a Mitigation Measure

Given burgeoning global emissions, there is a narrowing chance of limiting global warming to 1.5 °C by 2050, and therefore a case for pricing emissions as the fastest way of achieving change (Rogelj et al., 2018) particularly with best practice design and the support of complementary mechanisms. Pricing may be direct, like a tax, or less direct, via policies, regulations or carbon trading schemes. However, no carbon pricing scheme stands alone. It needs a supportive, political, institutional and fiscal environment, whilst the scheme itself requires workable, adaptive design features, with compensatory mechanisms to alleviate undue social impact in transitional circumstances. Pricing alone is also a blunt policy instrument, which requires complementary mechanisms, for sectoral adjustments, for example, for research and technological development to facilitate change, for information and communications management to address the socio-political challenges, and to address equity issues. Carbon pricing schemes are not easily comparable in climate leadership terms. They are necessarily context specific, by design, and in terms of scope, from narrow to comprehensive, in terms of the rate at which they price carbon, and in terms of their operational style, so they vary enormously, which is a limitation for the linking of carbon trading schemes bilaterally, regionally and beyond.

By 2023, only 23% of global greenhouse gas (GHG) emissions were covered by carbon pricing schemes (Twidale, 2023), but it is worth revisiting their pros and cons. Baranzini et al. (2017) see the main purpose of carbon pricing as achieving environmental effectiveness at a relatively low cost, which enhances the social and political acceptability of climate policy. They argue that the impact is immediate, as firms and consumers adjust to the added carbon price, which ideally would cover a broad range of emitters, and link with schemes abroad to minimise the cost of pollution control. Solutions would be sought for less carbon intensive technologies and activities, so carbon pricing would stimulate investment in innovation. However, Pearse and Böhm (2014) identify numerous problems including 'ineffectiveness, weak regulation and implementation, instances of fraud, little to no emissions reduction and major legitimacy issues for governments and the private sector' (Pearse & Böhm, 2014, p. 325). Amidst the many shortcomings of carbon pricing are that offset loopholes allow developed countries to push emissions reduction onto lesser-developed and developing countries with only lip service paid to sustainable development, and even worse, that compensatory funding to fossil fuel industries can deliver windfall corporate profits (Pearse & Böhm, 2014).

Nevertheless, the UNFCCC explains that a carbon price creates an incentive for emitting less. In theory, it shifts the cost of emitting carbon, and the socio-economic and environmental damages, impacts and vulnerabilities that this causes, away from the public, and the public purse, to the source of the problem which is the GHG producers. Again, in theory, the efforts of GHG producers to avoid paying this cost will spur changed processes and technological innovation, although in practice many a carbon pricing scheme has been distorted by overcompensating GHG producers. The theoretical benefits of carbon pricing range therefore from cost-effective and flexible pollution reduction to the pursuit of more sustainable development and a lower carbon economy, to innovation and job creation in an emerging lower carbon economy. Finally, revenue generated by carbon pricing must be allocated equitably, ideally as direct compensation to low-income affected householders, as assistance to producers in acquiring less carbon intensive processes or in transitioning out of the industry, and as funding for clean energy research and innovation. So, in theory, at least carbon pricing revenue would also promote an energy transition to a greener economy (UNFCCC, 2023).

However, the practice of carbon pricing demonstrates the gap between pricing theory and pricing in practice, and thus in the efficacy of reducing emissions (Stokes & Mildenberger, 2020). We should note that the 'carbon pricing mechanism' (CPM) typically refers to a carbon tax, to create a price signal that is felt across an entire economy, which incentivises a move away from carbon-intensive production and which results in a total reduction of emissions (UNFCCC, 2023). Or it can refer to a carbon trading scheme, otherwise known as an emissions trading scheme (ETS). An ETS is a less direct, more complex measure. It caps the total allowable emissions, then allocates one permit per ton of allowable emissions and facilitates the buying and trading of permits typically on carbon markets or exchanges. In this 'cap and trade' system, a company that emits less tonnage of emissions than it holds permits for can trade its 'excess to needs' permits to a company that needs them to cover its own emissions. Occasionally a carbon pricing scheme may include both tax and trading elements (ICAP, 2019). Less typical carbon pricing schemes include Australia's entirely ineffective emissions reduction fund (ERF), whereby the government purchases credits using taxpayers' funds from emissions reducing projects that purport, for example, to avoid deforestation, to regenerate native forests and to combust methane from landfills (ANUCOL, 2022).

Despite the logic that carbon pricing is the least cost, most efficacious mitigation policy tool, there has been an uneven embrace of the practice despite countries such as Finland, Poland, Sweden, Norway, Denmark and Slovenia all adopting carbon taxes in the 1990s. Following the early introduction of carbon taxation in Nordic and East European countries, the European Union established the first major regional ETS (the EU-ETS) in 2005. It is the most mature, complex regional carbon market to date and was followed by the establishment of the Regional Greenhouse Gas Initiative (RGGI) in North America in 2009. By 2023, there were 73 implemented carbon pricing mechanisms, in 39 national jurisdictions and 33 subnational jurisdictions, with revenue approaching US$100 billion, and few instances of being wound back or delayed because of the energy crisis caused by Russia's invasion of Ukraine (World Bank, 2023c, pp. 8–9). However, whilst these mechanisms cover 23% of global GHG emissions, less than 5% of emissions are covered by a carbon price at or above the price range that is needed to be on track to meet *Paris Agreement* reduction targets (World Bank, 2023c, p. 20).

Carbon Pricing in EA&P – A Stocktake

In Chapter 1 we saw that, in 2022, EA&P countries were responsible for 40.53% of global GHG emissions (Table 1.2). APEC has found that, in 2021, in the broader Asia Pacific, there were 35 carbon pricing schemes, but that these only cover 5.5% of global GHG emissions. The carbon prices set by these schemes are exceedingly low, at US$10/t$CO_{2e}$ for more than half of the emissions covered. And it found that the carbon price required to align with the goals of the *Paris Agreement*, namely US$75–100 per ton, has been applied to only 5% of the emissions covered by these schemes (APEC Secretariat, 2022). Global GHG emissions need to urgently reduce by 6–8% per year, to achieve an overall cut of 40–45% by 2030 to align with *Paris Agreement* targets (Climate Analytics, 2023). There is a discrepancy, therefore, between the significant contribution of the Asia Pacific to GHG emissions, the scale of emission reduction that is required and the scant impact so far on reducing emissions by the carbon pricing schemes in the region (Chapter 4). For these schemes to have an impact, their design needs to be efficacious, with broad coverage of the sources of emissions, and a cost per ton of emissions closer to the recommended US$75–100 that will trigger a shift away from fossil fuels. The goal of these schemes should be supported by complementary policies.

Table 6.1 identifies carbon pricing and trading schemes in EA&P, rather than in the broader Asia Pacific, and includes schemes that have been implemented, that have lapsed or have been abandoned, and that are under consideration with anticipated launch dates. There are proposed schemes scheduled for implementation in Japan (ETS), Indonesia (ETS) and Viet Nam (ETS), as well as an economy wide carbon tax proposed for Indonesia, and economy wide ETS schemes that are only under consideration at this stage in Thailand, ROC Taiwan (Taiwan) and the Philippines (APEC Secretariat, 2022, p. 9). The World Bank (2023c, p. 26) notes that Japan's Ministry of Economy, Trade, and Industry has presented plans that could see a national ETS launched in Japan in 2026; and that, in 2023, Taiwan legislated for a carbon tax on large emitters, as well as a carbon border adjustment measure (CBAM) for carbon-intensive imports, however without any detail at this stage. In 2022, the EU adopted a CBAM which has now commenced its 2023–26 non-compliant pilot phase. It adds a carbon price/tax to goods imported into its countries from its trading partners whose countries have no carbon pricing (Errendal et al., 2023, p. 22). The EA&P countries of China, South Korea, and potentially Australia, are amongst the EU's largest trading partners and the most likely to be impacted, if they have no scheme that satisfies the EU definition of carbon pricing. Additionally, EA&P countries may consider adopting their own CBAMs. China now has an ETS scheme. It has asked the WTO to review

Table 6.1 Carbon pricing in EA&P

Year	*Country and region*	*Carbon tax*	*ETS*
2005–13	Japan		Pilot JVETS (voluntary)
2008–11	Japan		ETS (pilot economy wide)
	New Zealand		NZETS (economy wide)
2010	Tokyo, Japan		Tokyo ETS
2011	Saitama, Japan		Saitama ETS
2012	Japan	Carbon tax	
2012–14	Australia		Carbon Pricing Mechanism+
2013	Tianjin, China		Pilot ETS
	Shenzhen, China		Pilot ETS
	Shanghai, China		Pilot ETS
	Beijing, China		Pilot ETS
	Guangdong, China		Pilot ETS
2014	Hubei, China		Pilot ETS
	Chongqing, China		Pilot ETS
2015	South Korea		K-ETS (best practice economy wide)
	Thailand		Pilot T-VER (voluntary)
2016	Australia		ERF Safeguard Mechanism
	Fujian, China		Pilot ETS
2019	Singapore	Carbon tax#	
2021	China		ETS (the world's largest carbon market)
	Indonesia		Pilot ETS (voluntary)
2023	Indonesia		ETS (for coal power plants)
2024*	Indonesia	Carbon tax#	
2025/26*	Viet Nam		Pilot ETS (to be launched)
2023/24*	Japan		Trial nation-wide ETS
2026/27*	Japan		Commence nation-wide ETS

Sources: APEC Secretariat (2022); World Bank (2023d).
Notes:
*Planned or under consideration.
+Disbanded.
#Economy wide.

the EU's CBAM, which Australia has argued is a threat to global growth (Di Sario & Leali, 2023).

EA&P Carbon Pricing in Coal Challenged Countries

The most coal challenged countries in EA&P in terms of coal production, consumption and exports (Table 2.1) are China, Australia and Indonesia. China has moved to an economy wide ETS after extensive regional trials, Australia was the first country in the world to adopt carbon pricing, only to then abandon it (Crowley, 2017), and Indonesia will introduce an ETS in 2023 and carbon tax in 2024. China's regional ETS pilots began in 2013. Its ETS was launched in 2021, and, although it only covers coal and gas plants, these contribute to 10% of global emissions. Coverage will expand to other high-emitting industry sub-sectors, but the challenge will be to overcome the limitations and design flaws of the current scheme and to bolster its ability to drive national decarbonisation (CAT, 2024). Australia has retained a historically ineffective 'safeguard mechanism' which may transition to a type of ETS because from 2023 it now sets an emissions baseline for over 200 facilities (World Bank, 2023c), but loopholes may see the scheme subsidise highly polluting fossil fuel projects (Verstegen, 2023). In 2023, Indonesia launched an ETS for the power sector to cover 99 coal-fired power plants and 26% of its emissions (World Bank, 2023c).

However, whatever domestic emissions reductions Indonesia and Australia might make from carbon pricing are insignificant compared to the global warming that results from their coal exports as the world's largest and second largest coal exporting nations, respectively (Table 2.1).

The most challenged countries in EA&P in terms of their ranking as coal importing/coal consuming nations are China (1st/1st), Japan (3rd/5th), Taiwan (4th/14th), Viet Nam (5th/12th) and Malaysia (9th/19th) (Table 2.1). The domestic emissions reductions that might be achieved by carbon pricing in these countries are undermined by their coal imports and consumption. Whilst China is moving towards nuclear energy and adopting renewable energy at record rates, and despite its nation-wide ETS covering the power sector, in 2022 fossil fuel power still accounted for 56% of its domestic capacity (EIA, 2022; see Chapter 7). And despite its experimentation with ETS schemes, adoption of a carbon tax and encouragement of decarbonisation in key sectors, Japan is wedded to voluntary action, fossil fuels, carbon capture, and economic growth and energy security over climate action (CAT, 2024; Abe & Arimura, 2022). Taiwan has been considering an ETS and a carbon tax since 2017 but has only introduced mandatory emissions level reporting, and a carbon fee or levy system, for large emitters. Viet Nam is, however, fully committed to a pilot ETS in 2026, to be operational by 2028, and to a declining ETS emissions cap that aligns with the targets in Viet Nam's Nationally Determined Contributions. Malaysia is at the exploratory stage of carbon pricing but has established a carbon exchange (World Bank, 2023c; see also Chapter 4).

Carbon Pricing in Philippines, Thailand and Singapore

The Philippines is a lower middle-income country at high risk from climate change (17th) (Table 5.1), and on the cusp of ranking in the top 20 coal producing (22nd), consuming (23rd), importing (10th) and exporting (12th) nations in the world (Table 2.1). It has yet to adopt a carbon tax or ETS; however, it has adopted a fuel exercise tax, which the OECD calls an implicit form of carbon pricing, and which covers 52.4% of GHG emissions. It has also stopped subsidising fossil fuels (OECD, 2022b). Thailand is an upper middle-income country with a relatively challenging reliance upon fossil fuels with world rankings of coal production (24th), coal consumption (18th), coal importation (11th) and coal exports (35th) (Table 2.1). Since 2015, it has been experimenting with voluntary carbon pricing, its T-VER, and has expanded its coverage significantly, and, in 2022, it established a carbon exchange. The very low carbon credit value in Thailand is expected to rise in line with global demand by linking the T-VER to the international voluntary GHG reduction market (Econ Digest, 2022). Singapore has negligible reliance upon or production of fossil fuels (Table 2.1) and negligible per capita emissions (Table 1.2) but will increase its carbon tax, which it established in 2019, to $5 per tonne in 2024 and $80 by 2030. It has set best practice eligibility criteria enabling companies to offset up to 5% of their carbon tax liability overseas (Tan, 2023).

South Korea Carbon Trading

South Korea is an upper income country, with a high level of preparedness for climate change, however, with a reliance upon imports (fossil fuels – crude oil, coal and liquified natural gas) for about 92% of its energy sources (Statista, 2023b). It is committed to net zero emissions by 2050; however, its slow and less than stringent emissions reduction efforts have been rated ‘highly insufficient’ by CAT. It is deprioritising renewables and is not *Paris Agreement* compliant, with no plans for phasing out coal or gas by 2030. Coal comprises about 35% of its power generation (CAT, 2024; see South Korea). But South Korea established the first nationally mandated cap-and-trade carbon trading scheme in East Asia in 2015, the Korean Emissions Trading Scheme (KETS), which now covers approximately 74% of national GHG emissions. It has the second-largest carbon market in the world after the EU’s ETS. The KETS was preceded by the 2010 Target Management System for Greenhouse Gases and Energy (TMS), an internally focused command-and-control system with no trading. Firms required to report and reduce emissions under the TMS switch to the KETS once they emit over a certain level of GHG emissions over a designated period. Nyonho et al. (2023) find that, between 2011 and 2017, the KETS did not significantly reduce emissions but that it may have improved the aggregate efficiency of energy use in the energy and manufacturing sectors.

New Zealand and the Pacific

New Zealand was the first country in EA&P, and one of the first outside Europe, to adopt carbon pricing, the NZETS, in 2008. The NZETS is a unique scheme for its comprehensive coverage and serves as the country’s main climate policy tool to achieve its legislated goal of being the first carbon neutral country in the Pacific region. Approximately half of the country’s domestic emissions are covered, namely those from the power, industry, building, transport, domestic aviation, waste and forestry sectors, with plans to be the first nation to cover agricultural emissions. However, the effectiveness of the NZETS suffered from initially failing to set limits on the use of international offset credits for domestic compliance, which then put at risk the goals of domestic emissions abatement and resulted in a shift to an internal carbon market only from 2015 (World Bank, 2023c; see New Zealand). The scheme has also been widely criticised for failing to reduce emissions, with New Zealand's *He Pou a Rangi* Climate Change Commission's advice that carbon pricing alone will not drive the low-emissions transition and that an ETS needs complementary policy as we have also argued above (Hall, 2021).

Fiji is the only other Pacific country to currently be contemplating the introduction of an ETS as a mitigation measure. Section 43 of Fiji's 2021 Climate Change Act empowers the Minister to pursue a carbon pricing mechanism, including an ETS, and prescribe fees and charges payable on the volume of GHG emissions above a prescribed level (Sloan, 2021). Otherwise, Samoa and the Federated States of Micronesia are levying fuel taxes to suppress fuel consumption, and fuel and energy efficiency taxes as a form of carbon pricing, whilst the Marshall and Solomon Islands have been advocating for a global carbon tax on shipping and aviation (Gerretsen, 2023).

Leading Carbon Pricing in EA&P?

There are now nearly two decades of experience with carbon pricing in EA&P, with a growing wealth of experience in various mechanisms at regional and national levels over the last decade but no way of determining the leading schemes given comparability issues. Those countries which contribute most to EA&P and global GHG emissions, namely China, Japan, Indonesia, Australia, Korea, Viet Nam and Thailand, all have, or plan to have, carbon pricing. However, there are no clear climate leaders in terms of adopting efficacious carbon pricing. In terms of pioneering carbon pricing, Japan and New Zealand were leaders for their early embrace of ETS pilots and schemes. And in terms of entrepreneurial leadership, Korea is currently leading in developing the largest market. Singapore is also leading in terms of the integrity and ambition of its carbon tax. And both Viet Nam and Fiji are regional leaders for their respective plans to adopt ETS schemes that align closely with their NDC plans. And yet such leadership has not impacted and reversed the burgeoning GHG emissions in the region because EA&P schemes lack breadth of coverage, efficacious implementation, sufficient incentives for change and robust carbon pricing. Too few carbon pricing schemes in EA&P cover too few GHG emissions, at too low a cost, and with too little ambition at present to significantly reverse emissions.

This parallels the situation globally and bodes poorly for carbon pricing-based action. However, APEC notes that if a universal price on emissions in the Asia Pacific could be 'set at US$25 per ton' and 'introduced collectively and gradually in the region over the next 10 years', emissions could be cut by 21% by 2030, exceeding the region's targets under the *Paris Agreement* by

8% (APEC Secretariat, 2022, p. 6). Nevertheless, the gap between pricing theory and practice at this stage remains pronounced. The problem is a failure to learn:

> ...environmentally effective and economically efficient carbon emission trading systems have eluded both the international community and the European Union, and in practice have arguably increased emissions by artificially prolonging and legitimizing reliance on fossil fuels.
>
> (Harris, 2014, p. 755)

Why then, Harris asks, would the Pacific Rim countries, Australia, China, Japan, New Zealand and South Korea, be any different? Rather than learn from the previous failures of the *Kyoto Protocol's* carbon market and the EU's ETS, the ETS schemes adopted by these countries have replicated the unsound design characteristics that made the earlier schemes ineffective. ETS systems are thus intrinsically flawed and not suited to emissions reduction. EA&P countries have a further decade's experience in carbon markets, in Australia's case by abandoning its carbon pricing, but shortcomings in ETS schemes still feature in their design. As Harris finds, ETSs require complex legal, institutional and technological investment, and stringent measuring, monitoring, and verification systems without which they are predisposed to failure. But Harris suggests that Pacific Rim schemes are marred by political decision-making – to allocate free permits, to allocate permits based on flawed, often self-reported, information, and to accept offsets as substitutes for emission reductions. This disincentivises emissions reductions and prolongs reliance upon fossil fuels (Harris, 2014, pp. 769–70, 792).

Regional EA&P Carbon Market?

Notwithstanding the failures of ETS and carbon pricing schemes over the last 20–30 years to reduce emissions, there is still an appetite to adopt them, indeed, to create linkages between ETS schemes via regional and international carbon markets. The theoretical benefits include reducing the costs of climate action, fostering regional trade, promoting carbon market stability, sharing resources, experience and expenses, and curtailing carbon leakage (Duggal, 2020). In August 2023, ASEAN Economic Ministers released the ASEAN Strategy for Carbon Neutrality (SCN) which, at the regional level, will complement Member States' efforts to meet their respective NDCs under the *Paris Agreement.* Not only is this strategy environmentally beneficial, the Ministers suggest, but the Member States also expect it to generate between a US$3.0 and $5.3 trillion GDP value-add by 2050, attracting a substantial US$3.7 to US$6.7 trillion in green investment and unlocking between 49 and 69 million additional jobs for the region. One of the eight key strategies of the SCN is to enhance the interoperability of carbon markets (ASEAN, 2023c). However, given the lack of maturity and emission reduction success thus far in EA&P ETS schemes, lessons must be learnt on market linking from further abroad. Harmonisation of schemes prior to regionalisation or bilateral linking is critical but will be a challenge in the EA&P context where there are starkly differing governance and climate readiness capacities between developed, lesser-developed and developing countries (Nakamura & Crowley, 2020; Table 5.1).

It remains to be seen how a regional carbon market may work for either Asia Pacific or the EA&P, but the options include EA&P domestic markets fully, partially or indirectly linking with each other, or linking one way only without reciprocating (APEC Secretariat, 2022, p. 43). Bilateral linkages may also either evolve into a regional ETS or a regional ETS may emerge by design. Some EA&P countries have already proposed to link, or have already linked, with the EU and/or the North American RGGI. South Korea is pursuing emission reducing partnership agreements with 17 countries including Viet Nam, and the major GHG emitters Brazil, India and Indonesia under Article 6.2 of the *Paris Agreement* (Reklev, 2022). This article:

> … outlines the possibility of cooperative approaches and the transfer of Internationally Transferrable Mitigation Outcomes (ITMOs) between different actors, including countries and private sector companies, through bilateral agreements.
>
> (Soezer, 2022)

Article 6 was agreed in 2015 as a means of adopting instruments that would facilitate international cooperation to achieve GHG emissions reductions through carbon markets. It provides a framework for the bilateral or multilateral linking of carbon markets, in the context of robust and verified accounting processes that reduce emissions, that do not double count emissions reductions and that contribute to sustainable development (Duggal, 2020). It promotes higher ambition in emissions reduction, by allowing greater flexibility including voluntary cooperation between Parties and the transfer of ITMOs. The first agreements under this article involved EA&P countries. Vanuatu partnered with Switzerland and the United Nations Development Programme. Thailand partnered with Switzerland, and Singapore has signed several MoUs with crediting mechanisms in the context of supplying credits for its carbon tax (World Bank, 2023c, p. 47). Whether authentic emissions reductions will be delivered is not yet clear. The EU and developed countries would be credited via ITMOs with emissions reductions in EA&P at a cheaper price per ton than in their own countries. Rosales (2023) observes that ASEAN countries that 'are rich in biodiversity, forests and renewable energy sources such as hydro, solar and geothermal' have ambitions to become 'hubs for trading carbon credits', and that investments in these areas could generate a significant number of ITMOs that would ensure partnering countries meet their own NDC commitments at greatly reduced cost.

Regional Carbon Trading Design

The impetus for more urgent polycentric global-regional-national climate action in EA&P is clear in the evidence to date that individual nations alone are failing to generate the scale of the emissions reduction required by the *Paris Agreement*. In terms of developing regional carbon trading (RCT) as part of the solution to the climate change problem, the EA&P geographic, institutional and cultural context is more challenging than the geographically more concentrated, institutionally more integrated and culturally more regulated context in Europe. Existing EA&P ETS schemes, as well as those that may be adopted, need to be designed to harmonise, whilst retaining relevant domestic differences, as ETS linkages are attempted in bilateral, regional and global contexts. The prior experience offers lessons in facilitating ETS linkages in terms of developing a framework of harmonised market rules, information sharing, institutional arrangements and legislative structure that is credible, transparent and fair to its participants (see Duggal, 2020). It also shows how an EA&P regional carbon market could be expanded to absorb new participants, with Norway, Iceland and Liechtenstein joining the EU ETS in 2007, and Switzerland in 2020, and the Californian and Quebecois markets integrating in 2014 (World Bank, 2023c).

We have previously considered the design challenges facing the development of RCT in EA&P. RCT in EA&P would aim to reduce emissions and lessen the vulnerability of the region to climate impacts, in a manner that is integrated with development aims. It would facilitate harmonisation between domestic ETS schemes but allow for differentiated capability and capacity in emission reduction efforts between developed, lesser-developed and developing countries. It would operate to reduce emissions, but also to promote sustainable development, with the aims of diminishing fossil fuel driven economic growth in the region and also of generating the capital to invest in innovative low and zero carbon technologies. Lesser-developed and developing EA&P nations would understandably have concerns about hampering their economic growth by acting to reduce their domestic emissions, including by developing and joining any carbon pricing schemes. Their establishment of carbon markets and exchanges, and their issuance of ITMOs or other carbon crediting mechanisms, would need to be of an independently verified standard to balance their capital raising ambitions with the broader goals of emission reduction, sustainable development and greener economic growth (Nakamura & Crowley, 2020).

Complementary Policy and Best Practice Design Features

Given the nascent status of carbon pricing in EA&P, determining leading efforts is premature, but these would learn from previous experience and adopt elements, as appropriate of best practice design. Furthermore, best practice carbon pricing would sit within a complementary set of market and

Table 6.2 Complementary policies: types of instruments for regulating GHG emissions

Non-market-based (direct limitation of harmful anthropogenic impact on the ecosystem)
∇ _Technical regulation
∇ _Resource expenditure rate (gasoline consumption standards, building energy efficiency standards, etc.)
∇ _Best Available Technologies (compiling a list of technologies that are both technologically accessible and best meet the goals of environmental protection. Technologies from these directories are gradually becoming mandatory for companies)
∇ _Voluntary environmental agreements between industry and government
∇ _Quantified emission limitation
Economic or market
Influencing incentives for emissions reduction:
∇ _Carbon tax
∇ _Emissions trading system
∇ _Subsidies for emissions reduction (including subsidies for the use of renewable energy sources and other low emissions energy sources)
∇ _Fossil fuel subsidy reforms (balanced by the requirement to provide those in need with essential energy services)
∇ _Crediting schemes
Influencing incentives for the production or consumption of emissions intensive products:
∇ _Tax on emissions intensive products
∇ _Low emissions products subsidies

Source: APEC Secretariat (2022, p. 5).

non-market-based climate policy tools (Baranzini et al., 2017; Table 6.2). The New Zealand ETS, lauded for its early establishment, has also been criticised, for example, for this not being the case (Hall, 2021). The IMF explains that:

> …(m)itigation measures can be packaged to meet different policy objectives and implementation challenges. In practice, a combination of mitigation measures is often adopted in the (Asia-Pacific) region. For example, Singapore has both a carbon tax and a feebate system, together with a fuel excise tax for the transport sector, and Japan has both a carbon tax and regional ETSs, while China, Korea and New Zealand have regulations on air quality and fuel standards to complement their ETS systems. The mix of measures depends on implementation challenges and specific policy objectives.
>
> (IMF, 2021c, p. 15)

Complementary policies also place the actions of self-interested market players within a regulatory landscape amidst a set of reinforcing policies specific to domestic circumstances (IMF, 2021b). The mix of these will likely include fiscal and structural reforms, public procurement, carbon pricing, stringent standards, information schemes, technology policies, fossil-fuel subsidy removal, climate risk disclosure, and land-use and transport planning (OECD, 2017). The finance generated by carbon pricing, from an ETS scheme and/or from carbon taxes, can help to offset the cost of complementary measures, mitigate adverse cost impacts upon households, industries, firms and workers and to provide green financing for research and development and technological innovation, for upskilling the workforce and for infrastructure. There are also sectors for which carbon pricing is not feasible and regulatory policy is required, and others where decarbonisation would be extremely difficult without the support of complementary policies (IMF, 2021b).

To counter the variation in complementary climate measures, the IMF has identified the core features of best practice design which would be adopted by climate leading countries. These are *Fairness* (ensuring that the polluter pays); *Alignment of policies and objectives* (including complementary measures); *Stability and predictability* (including steadily rising carbon prices over time); *Transparency; Efficiency and cost effectiveness* (to help increase ambition); and *Reliability and environmental integrity* (to ensure real environmental benefits) (OECD, 2017, p. 5). Regional bodies within East Asia and the Pacific could adopt a regional emissions reduction target, and set its parameters, including regulatory and accounting mechanisms, consistent with the aims of the *Paris Agreement.* An EA&P RCT scheme could achieve harmonisation through complementary regional and domestic legislative frameworks. The allocation of emission permits and credits could extend to the allocation of enforceable penalties for breaching emission caps, with penalties levied for breaching the requirements of the scheme. The scheme could build in sufficient flexibility to respond to crisis, change, corruption, expansion and so forth, with additional flexibility to enable further linkages as the scheme evolves. Market benefit could be legislatively articulated to ensure that finances raised build the region's mitigation and resilience building capacity, and ensure market stability (see Nakamura & Crowley, 2020). An EA&P scheme should prioritise mitigation and resilience building over revenue raising, prolonging fossil fuels, and offsetting.

Conclusions

Although we include the adoption of carbon pricing as a criterion for assessing regional climate leadership, the paradoxes of carbon trading are clearly many, most blatantly, where it prioritises revenue raising, where it fails to reduce emissions and where it diverts or displaces the goals of the *Paris Agreement.* Whilst the GHG emissions covered by carbon pricing have fallen by c.0–2% per year since the 1990s, ETS and carbon tax prices hit record highs in many jurisdictions in 2021–22 (Errendal et al., 2023, p. 9). It is also unclear whether carbon trading incentivises an exit from fossil fuels and a transition to renewables, with many critiques arguing that the reverse can be true, that fossil fuel longevity can be built into schemes, which can be undermined by fossil fuel subsidies. In theory, carbon trading is the least cost, fast results climate policy tool, but in practice its design can be distorted, its implementation flawed and its mitigation efficacy can be diminished. Worse may be the danger that developed nations trade away their NDC emission reduction obligations through carbon markets in EA&P under the auspices of Article 6 with an impact on reducing emissions. However, the literature is replete with best practice carbon pricing and leading emissions trading policy design, which we have distilled and discussed above (Nakamura & Crowley, 2020). The challenge, for any EA&P nation aspiring to climate leadership, would be to embed carbon trading within a suite of complementary policies and to design in features that ensure it is an effective mechanism for reducing emissions.

7 China, Regional Leadership in the EA&P?

Introduction

In the previous chapters, we have seen that East Asia and the Pacific region (EA&P) is responsible for 40.53% of global greenhouse gas (GHG) emissions, with China responsible for the bulk of those (29.16%). China's emissions have also grown 203% since 1990 (Table 1.2). We have also observed that China has emerged as a dominant regional power (Dent, 2008), but with a contradictory record whereby its growth, security and territorial influence all play a role in its climate policy. More recently, China's economic expansion and political reach have increased security and trading tensions between China and EA&P countries, although China remains engaged in bilateral (Table 4.4) and regional dialogue (Table 2.1). We have seen that China has sought to enhance its regional influence by expanding its aid to Asia Pacific countries, including aid for building climate resilience predominantly but not exclusively through its expansionist Belt and Road infrastructure initiative. In the *Paris Agreement* era, China has pivoted decisively towards regional and global climate leadership (Hurri, 2020), which we will consider here, and is making a rapid transition to a lower carbon economy, but its continued reliance upon fossil fuels remains a major obstacle to overcome.

China has been the world's fastest-growing country for the last three decades and has largely driven the quadrupling of wealth between 2000 and 2020 in EA&P. China's GHG emissions grew by 75% between 2005 and 2019, three times the rate of the rest of the world (Erbach & Jochheim, 2022). It is now the world's second largest economy, its largest GHG emitter and its second most populous country (World Bank, 2024b). It ranks second within the top 20 countries globally in terms of cumulative (billion tonnes of CO_2 from fossil fuels, cement, land use and forestry) emissions from 1850 to 2021 (Carbon Brief, 2021). It has the world's largest number of coal power plants and also the greatest capacity for renewable energy (IRENA, 2024; WPR, 2024), as well as the third greatest capacity for nuclear power (IAEA, 2024). It is the world's fifth largest oil producer and the second largest consumer of oil (Energy Institute, 2023). In Chapter 2, we saw that China has only 13.3% of proven coal reserves globally, but that it has 50.8% of total global coal production and 53.8% of total global coal consumption (Energy

DOI: 10.4324/9781003170389-7

Institute, 2023). Rather than ramping up coal production as China has done in recent years (Hawkins, 2023), as a climate leader it would reverse its reliance upon coal that could otherwise derail the global objectives of the *Paris Agreement.* China's structural power as an emergent superpower and its climate policy actions have global consequences, as we shall see.

China's Historical Role in Global Climate Change

From Kyoto to the Paris Agreement

China's role in responding to climate change has evolved significantly since the early 1990s. Initially, it was an early responder in 1990 for forming a coordinating committee, the National Climate Change Coordinating Group, to integrate the efforts of its various government agencies. It has been a 'non-Annex I' developing country party to UNFCCC efforts, processes and agreements, including ratifying the *Kyoto Protocol* and the *Paris Agreement* ever since the 1992 Rio Earth Summit (Sandalow et al., 2022). Wu (2023) explains that China in these early years saw climate change very much as an issue of foreign affairs, but that this stance would evolve over three distinct, differing periods in the history of China's climate policy efforts and engagement from periods of: being a defender of development rights (1988–2006), to being an active follower and participant (2007–15), and ultimately to being a global leader (Post 2016). Wu observes that initially, China's emissions were eclipsed by developed countries and were only a fraction of the United States' emissions. So, China initially engaged in UNFCCC processes as a means of securing its right, as a low-income, developing country, to defer emissions reductions and prioritise growth underpinned predominantly by coal-fired energy.

However, having secured its common but differentiated responsibility to act on climate change, and its right to enact no climate policies, Liu et al. (2023) observe that, by 2006, China's emissions had skyrocketed, as China rapidly developed and overtook the United States to become the world's largest annual emitter of GHGs. Climate change and emissions reductions then became very much a domestic as well as an international priority in the period that Wu (2023) identifies of China being an active follower and also an active participant in terms of global climate change efforts (2007–15). This was a period of policy catch-up and climate institution building which would nevertheless struggle to deliver outcomes by 2015, Wu suggests, but which laid the foundation for future accomplishments by establishing climate policy goals and administration. Significantly, a National Climate Change Program was released in 2007 as China's first climate policy, the Department of Climate Change was established in 2008, voluntary carbon emissions trading and an adaptation policy were adopted by 2013, and goals were set to achieve energy efficiency and to reduce carbon intensity and were integrated into China's Five-Year plans (Wu, 2023). It is important to note that China began integrating its climate policy and its economic development objectives.

There has been significant focus on Wu's post *Paris Agreement* period (Post 2016) but Sandalow et al. (2022) identify historically important lead in events and developments to this period. The release of China's *National Plan on Climate Change* (2014–20), for example, was an agenda setting moment, as was the 2014 joint statement by China and the United States, the world's two biggest emitters, on tackling climate change in which China, for the first time, committed to peak emissions by 2030. Whether a demonstration of China's climate leadership or pragmatic catch-up, this statement eventually smoothed the way to the international community achieving the 2016 *Paris Agreement* (Qi & Dauvergne, 2022). Following the 2014 joint statement, China then joined the international parties that were about to commit to the *Paris Agreement* by submitting its Intended Nationally Determined Contribution (INDC) in 2015 to the UNFCCC and pledging 'to achieve the peaking of carbon dioxide emissions around 2030, making [its] best efforts to peak early. It also pledged that by 2030, it would (1) lower carbon dioxide emissions per unit of GDP by 60%–65% from the 2005 level, (2) increase the share of non-fossil fuels in primary energy consumption to around 20%, and (3) increase its forest stock volume by around 4.5 billion cubic meters from 2005 levels' (Sandalow et al., 2022).

Post Paris Agreement

In 2017, when US President Trump pulled America out of the *Paris Agreement*, China was left a de facto global leader on climate change ambition if not actual achievements, whilst the Trump administration adopted an aggressive 'mindset of comprehensive confrontation' towards China (Arežina, 2019, p. 291). President Xi declared at the time that China was '[t]aking the driving seat in international cooperation to respond to climate change. China has become an important participant, contributor, and torchbearer in the global endeavour for ecological civilization' (Sandalow et al., 2022, Chapter 3). Despite its ambitions, China's emissions continued to rise. In the *Post Paris* period, with its emissions rising and US-China relations deteriorating, China's climate policy was more directly integrated into its economic, sustainability and national planning. Wu (2023, p. 4) details the significant administrative reforms that accompanied China's climate policy development, including its energy efficiency plans, its move towards carbon trading and its release of its updated 2030 and 2060 goals. By 2020, China was confident it would reach its goals to peak emissions by 2030 and reach carbon neutrality by 2060 before time; however, its efforts would not, China stressed, compromise its development trajectory, nor its economic or energy security (ASPI, 2023).

Following its 2020 goal setting, in 2022 China released updated climate policies as part of what it describes as 'a national strategy of proactively responding to climate change by giving it higher priority in the national governance system'. China's peaking emissions and carbon neutrality goals are described as integrated into its overall plans for promoting ecological civilisation, economic and social development. The focus of its climate policy efforts is upon achieving synergy between reducing pollution, controlling emissions and promoting a comprehensive transformation towards environment-friendly economic and social development (MOE&E, 2022, p.1). China's key policy platform for peaking emissions and achieving carbon neutrality was described in 2022 as '1+N', whereby '1' refers to its overarching 2021 emissions and carbon neutrality goal setting and policy context, and 'N' refers to its implementation schemes in key areas and sectors – energy, industry, transport, urban and rural development, and agriculture (MOE&E, 2022, p.1). The strategy, *China's Policies and Actions for Addressing Climate Change 2022*, strengthens the top-level design and integrated implementation of climate policy, with a national level coordinating group and coordinating groups established in all provinces, autonomous regions and municipalities.

Currently, China is implementing its 14th Five-Year Plan (2021–25), but its climate policy goals have been dented by the challenges of the Covid-19 pandemic, the global impacts of Russia's invasion of Ukraine and drought-induced hydro-energy insecurity issues. The Covid-19 pandemic was a massive social, economic and ecological disruption around the world, which, in China, was exacerbated by the country's adoption of the world's strictest Covid-Zero lockdown practice. This practice was only lifted after three years in early 2023 and had been intended to prevent infection and to minimise its transmission when it broke out, with prolonged lockdowns, not only of China from the outside world, but wherever there were outbreaks within the country. With economic and industrial activity dramatically slowed during the Covid pandemic, at times to a standstill, China's carbon emissions dropped overall, as did the rest of the world's, in a short-lived reprieve in the trend otherwise to globally accelerating emissions (Ronaghi & Scorsone, 2023). Conversely, China's post-pandemic stimulatory industrial and economic policies saw a huge surge in emissions, and in coal power construction and consumption, with expectations however that this will be temporary, given the 'staggering' post-pandemic investment in China's low carbon energy (Myllyvirta & Qin, 2023).

Russia's invasion of Ukraine has, at the very least, unsettled the global energy market, at best by accelerating the drive to renewable energy as a substitute for Russian oil and gas, and at worst by protective measures against excessive energy costs driving up the subsidisation of fossil fuels (Muta & Erdogan, 2023; Psaropoulos, 2023). Western sanctions levelled against Russian oil exports benefited China which leveraged reduced priced oil imports from Russia, directly and via Malaysia, saving 'EUR 18 billion over 18 months, or around 3.5% of the total value of crude oil imports' (Ishii et al., 2023). There is some concern that the International Energy Agency may be wrong, that China's pandemic and drought-driven acceleration in coal-fired power generation may not be temporary and that the primacy of national security may thwart its transition to renewables. In the wake of China's 2022 plan for accelerated climate change policies and action, the country was gripped by drought in its southern provinces that increased its reliance on coal for energy as its hydro-reserves plummeted (Kemp, 2023). The resultant increase in coal consumption in China may, Shaw (2024) argues, be prolonged into 2024 because of energy supply issues, issues with the viability of the nation's grid and a less than smooth adoption of increased renewables. Such issues in the uptake of renewables and obstacles in the phasing out coal are, however, typical in energy transitioning efforts well beyond China (Figure 7.1).

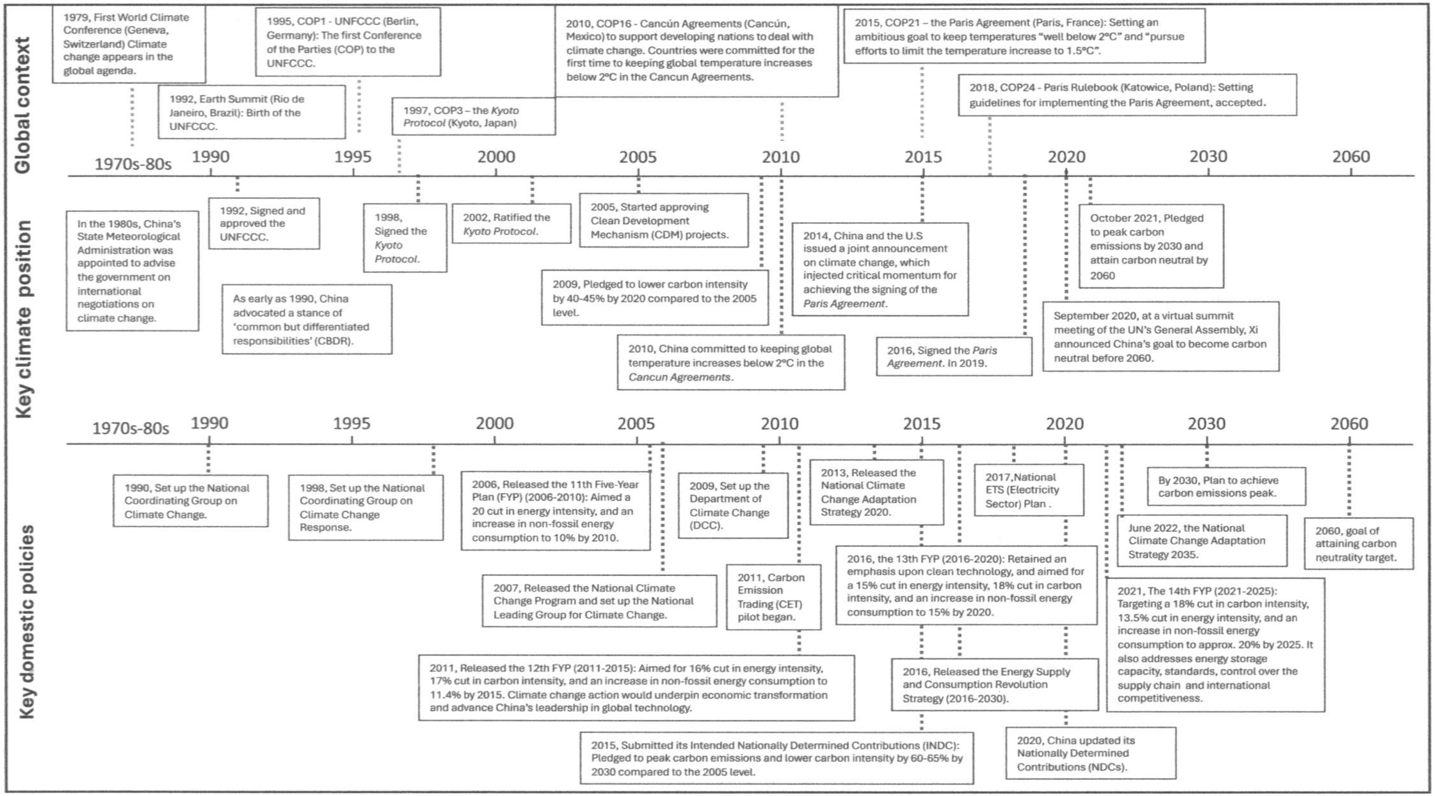

Figure 7.1 China's key climate change policies and positioning.

Sources: Carbon Brief (2023); EU (2022); Meidan (2020); World Bank (2022c); Wu (2023).

China's Climate Policy Performance 2024

China's Climate Change Performance Ranking

Through its travails over the last several years with the pandemic, the drought and energy and political disruptions from Russia's invasion of Ukraine, China has not updated its Nationally Determined Contribution to global efforts to meet the objectives of the *Paris Agreement.* Its policies remain as they were, and its climate leadership and lack of leadership in various categories (CCPI, 2024) are reflected by their implementation. In 2024, China is not tracking to meet its 2025 targets of increasing the share of non-fossil fuel energy to 20% and reducing carbon intensity by 18%, with the latter currently growing, since 2021, on an average by 3.8%. This is a direct consequence of corrective stimulatory measures put in place to deal with the pandemic, drought and energy crises that have boosted construction, renewable energy and the approval of coal-fired power (Hawkins, 2024). Indeed, despite President Xi's 2021 pledge at a US White House Leaders Summit on Climate to 'strictly control new coal-fired power generation', CREA (2024, p. 1) reports that 'approvals of new coal power plants increased 4-fold in 2022-23, compared with the previous five-year period of 2016-20'. And that '[s]ince the beginning of 2022, an estimated 218GW of new coal plants have been permitted. 89GW of this capacity had already started construction by the end of 2023, but [the construction of] another 128GW had yet to break ground'. The increasing not decreasing percentage of power generation from fossil fuels, since 2021, is also in part because the 2021–24 period has represented the last gasp opportunity to expand coal-fired power (CREA, 2024).

Under these circumstances, the Climate Change Performance Index ranks China's current overall performance as that of a climate policy laggard, with China holding on to its 2023 rank of 51st and performing poorly (a *very low* rating) in the GHG emissions and energy use categories, and only moderately (a *medium rating)* in terms of renewable energy and climate policy (CCPI, 2024). Despite this overall poor ranking and overall failure to be on track towards its 2025 goals, CCPI notes China's astonishing renewable uptake is a notable exception. The renewables sector and energy efficiency have both been experiencing robust growth, receiving a 'very high' rating in Table 7.1 below, which should deliver 1,200 GW of wind and solar power by 2025, rather than by China's target date of 2030 (CCPI, 2024). China's national climate policies are also ranked extremely highly by the CCPI (Table 7.1), and remain the guiding template for national, regional, provincial, and local policies and actions; however, their objectives have largely been displaced by the series of crises since 2021. Sun and Mi (2023) observe that the pattern of post-pandemic economic recovery in China from the latter half of 2020 has resulted in escalating, not declining, carbon emissions. They suggest that a greener, more resilient recovery from the pandemic will require the integration of China's climate policies with the promotion of low carbon patterns

Table 7.1 Climate Change Performance Index performance – China

Indicators – emissions 40%; RE 20%; Energy use 20%; Policy 20%	*Weight*	*Rating*	*Rank*
Category – GHG emissions	**40%**	**Very low**	**62**
Per capita – current level (including LULUCF)	10%	Low	47
Per capita – current trend (excluding LULUCF)	10%	Very low	62
Per capita – compared to well below 2 °C*	10%	Very low	54
2030 target – compared to well below 2 °C	10%	Very low	56
Category – Renewable energy	**20%**	**Medium**	**16**
Share of ⁺RE in energy use – current level (incl. hydro)	5%	Low	40
Renewable energy – current trend (excl. hydro)	5%	Very high	6
Share of ⁺RE in energy use (excl. hydro) compared to well below 2 °C	5%	Very low	32
⁺RE 2030 target (incl. hydro) compared to well below 2 °C	5%	Low	22
Category – Energy use	**20%**	**Very low**	**61**
Energy use per capita – current level	5%	Medium	30
Energy use per capita – the current trend	5%	Very low	65
Energy use per capita – compared to well below 2 °C	5%	Very low	53
Energy use 2030 target – compared to well below 2 °C	5%	Very low	58
Category – Climate policy	**20%**	**Medium**	**13**
National climate policy	10%	Medium	7
International climate policy	10%	Medium	23

Source: CCPI (2024).

Notes: ⁺RE – renewable energy. *Well below 2 °C (i.e. *Paris Agt* goal).

of consumption, enhanced investment in decarbonisation and the accelerated development of low carbon industries.

China's Climate Action Tracker Ranking

In November 2023, the Climate Action Tracker (CAT) rated China's climate policy performance as 'highly insufficient' in terms of its policies and actions, its NDC target of net zero by 2060 and what could be expected as its fair share of global climate policy and emission reduction efforts (CAT, 2024). Rather than peak in 2025 and then decline, CAT's assessment is that China's emissions will plateau, with the key challenge being that China's already rapid uptake of renewables must be sufficient to trigger its transition away from fossil fuels. The current burgeoning of fossil fuels as an energy market stabiliser, driver of economic recovery, and response to crisis is contrary to China's NDC and its climate policy commitments, but CAT (2024) observes that greater ambition is still possible in its climate, energy and sectoral targets. China will likely recommit to its climate targets when it updates its NDC potentially in late 2024, but possibly closer to its 2025 target deadlines and the release of its 15th Five-Year Plan (2026–30), which, CAT argues, must drive more ambitious decarbonisation. China's uptake of wind and solar power is world leading; however, as the rest of the developing world shed 23 GW of coal-fired power capacity in 2023, China's planned coal-fired power capacity of 126 GW more than dwarfed that achievement and threatens to derail the *Paris Agreement* targets. China is not alone in persisting with coal-fired power; however, its actions have an outsized influence on global climate goals with '[m]ore than half (59%) of the 45.5 GW of newly commissioned [coal fired power] capacity in China' in 2023 (GEM, 2023).

CAT assesses that China's policies and actions are inconsistent with achieving the ideal 1.5 °C warming limit and that if all countries had its 'highly insufficient' level of ambition, then global warming would be between 3 °C and 4 °C by the end of the century. Avoiding this would require a ramping up of the uptake of renewables, wind and solar, so that they replace rather than supplement coal-fired power, and to ensure that gas is relied upon only briefly. This is not just a matter of political will. Implementation issues must also be revisited. Financing the transition to a low-carbon economy is a particular challenge (Engels, 2018). The regulations governing China's carbon trading scheme, scope and operation need revisiting (Zhang, 2023). And in terms of administrative shortcomings, China is yet to legislate its climate policy, its policy instruments have not kept pace with its increased ambition, there is a vertical disintegration between the policy aims of national and local governments, and there is a lack of participatory governance (Wu, 2023). With China's current sluggish economy and accelerated carbon emissions, CAT's (2024) priorities for revising its climate policies are to: (1) control and reduce its fossil fuel dependence; (2) improve its climate targets by setting an absolute economy-wide peaking target for all GHG emissions; (3) raise the ambition of its energy-related climate (NDC) targets; and (4) accelerate the decarbonisation of high-emitting industry sectors.

China's Engagement Role in Climate Action

UNFCCC Driven Engagement

As Wu (2023) observes, prior to the *Paris Agreement*, China perceived climate change as a foreign affairs issue, and its early engagement at the international level on the formulation of the United Nations Framework Convention on Climate Change was driven by the desire to protect its developing country national interests. China ratified the UNFCCC in 1993 and assumed a leading role as the 'conduit' between Annex 1 developed and the G77 and other developing countries, not officially as a member of the G77, although at times as a G77+China partner. It argued forcefully for 'common but differentiated responsibilities', and for exemption as a non-Annex 1 developing country, from the emissions reducing measures in the *Kyoto Protocol*, and despite its domestic climate policy and institution building efforts, was seen internationally as an extreme climate laggard. But over time, and paralleling both its development efforts and the burgeoning emissions that resulted, China's positioning evolved from the defender of development rights (1988–2006) to an active follower and participant (2007–15) and to global leader (Post 2016) (Wu, 2023). In terms of engagement outside the UNFCCC, most significantly China has sought to engage with the United States, but it has also engaged with other developing countries, including in EA&P, through its Belt and Road initiative (Cain et al., 2022) and South-South Climate Cooperation Fund (Lewis, 2023).

Following the failure of the *Kyoto Protocol* to result in reduced emissions (Napoli, 2012), it was a historic, agenda setting moment when China and the United States, the world's two largest GHG emitters, made their 2014 joint announcement setting domestic climate goals and committing to co-operate. Their joint commitment to a strengthened and more effective post *Kyoto Protocol* agreement was critical to achieving the *Paris Agreement* and saw the Chinese delegation then become actively involved in COP21 negotiations in Paris in December 2015. The US-China climate change engagement deepened as China was one of the first countries to ratify the *Paris Agreement* the following year in a joint ceremony with the United States (Sandalow et al., 2022). The subsequent faltering of US-China relations on climate change followed the inauguration of conservative US President Donald Trump in 2017 and his aggressive, hawkish treatment of China, especially over Covid-19 which originated there (Arežina, 2019). Relations thawed between the two countries following the inauguration of Democrat President Joe Biden in 2021, who then appointed long-term climate campaigner John Kerry as the first US special presidential envoy for climate. By the end of that year, at COP26 in Glasgow, China and the United States were releasing another joint declaration, this time on enhancing climate action and cooperation into the 2020s (Sandalow et al., 2022). They committed to cooperating 'on a variety of climate change issues, including developing regulatory frameworks and environmental standards to reduce GHG emissions, encouraging decarbonisation and electrification of end-use sectors, and fostering the circular economy' (IEA, 2022b). They released another joint statement on 15 November 2023 ahead of COP28.

Belt and Road Initiative

China's Belt and Road Initiative (BRI), launched in 2013, is an audacious, globally significant funding initiative, currently involving over 155 countries, from low to high income, that invests in infrastructure critical to boosting trade, market access for China and global GDP. Ten years on, US$1.053 trillion has been invested globally by the BRI programme, including US$634 billion in construction contracts (much of it on infrastructure essential to transport and trade, but also until recently significant funding of offshore fossil fuel-based energy projects) and US$419 billion in other investments (Wang, 2024). Uren (2023) observes that the loans and debt restructuring offered by Chinese banks as part of the BRI initiative differ from those of the World Bank and IMF by being opaque, by carrying relatively high interest rates and by risking debt traps for developing countries. There have been three BRI forums, in 2017, 2019 and 2023, and, over time, there has been increasing concern at these that BRI infrastructure development has outpaced environmental governance, with no compliance mechanisms adopted for achieving sustainability goals, despite the 2017 establishment of a BRI International Green Development Coalition (Pike, 2019). However, in 2021, President Xi announced that China would build no more coal-fired power projects abroad (Carbon Brief, 2022), and by 2023 China's 'energy related engagement' was 'the greenest in absolute and relative terms in any period since the BRI's inception reaching US$7.9 billion' (Wang, 2024). Of the EA&P countries (Table 1.1), only Australia, Japan, the Marshall Islands, Nauru and Tuvalu are not participants in the BRI (GFDC, 2023; see Table 4.4).

EA&P Engagement and the South-South Cooperation Fund

In terms of general climate change engagement in EA&P (see also Tables 2.1 and 4.4), China and ASEAN recently released the *ASEAN-China Joint Statement on Enhancing Green and Sustainable Development Cooperation* at the 24th ASEAN-China Summit on 26 October 2021. This included a joint agreement to cooperate on climate change mitigation and adaptation, including through cooperation in the energy transition and in strengthening energy resilience (ASEAN-China, 2021). China and New Zealand signed a comprehensive strategic partnership on 1 January 2022, but these countries have long-standing prior environmental and climate-based cooperation including a 2017 joint action plan on climate change, a 2019 joint leaders' statement on climate change and a 2022 joint ministerial climate change dialogue (NZFA&T, 2023). China's recent climate change engagement in the Pacific includes climate change training sessions, supplies to tackle climate change, renewable energy cooperation and infrastructure support, and the launch, in April 2022, of the China-Pacific Island Countries Climate Change Cooperation Centre (PRC/FSM, 2024). As part of its climate change policy pivot in 2015, China also announced that it would not seek climate finance for itself from developed countries and that it would promote its own South-South Cooperation Fund (SSCF), rather than contribute to the UN's Green Climate Fund (Carbon Brief, 2023). This billion-dollar initiative promotes bilateral clean energy agreements, multilateral clean energy cooperation, the export of China's green energy technology and the delivery of clean energy infrastructure investment. Funding is delivered by bilateral partnership agreements and MoUs with developing countries and includes the '*Ten, Hundred, Thousand*' project consisting of 'ten low carbon demonstration projects in developing countries, one hundred climate change mitigation and adaptation projects, and one thousand climate change training projects' (Lewis, 2023, p. 202).

Is China a Climate Leader or Laggard?

China has pivoted remarkably on climate change since its 2014 joint statement with the United States. Its goals remain to achieve its 2030–60 emissions reduction and carbon neutrality goals in record time, with the world's sharpest drop of carbon emissions from carbon peak to carbon neutrality, but it faces greater challenges than developed countries, given that it must decarbonise whilst is still developing (Hurri & Kopra, 2023). Is climate leadership possible in these circumstances? China has acknowledged that it has a global leadership role to play, beyond that of a developing country, in setting aspirational emission reduction and carbon neutrality goals and in assuring that its policies are implemented so that its goals are more likely to be achieved. It held firm on its enhanced climate policy aspirations as a de facto global leader when the United States vacated the international arena rather than abandoning its plans, but it has struggled through the recent pandemic, global insecurity and drought crises to hold the line on reducing emissions. Unsurprisingly, China's climate reputation is now Janus-faced. It is both a climate leader and climate villain because its supercharged clean energy revolution is being undermined by the bourgeoning support of coal-fired power (Meidan, 2020). It is navigating change away from coal-fired power at home and abroad, but, Engels (2018) suggests, with as much concern for the negative energy security and health impacts of coal, and its limitations for transformational economic growth, as for addressing the impacts of climate change.

Leadership is, however, a nuanced concept. It may be self-declared, for example, when China's President announced that China was 'taking the driving seat' following the United States vacating global climate efforts. Indeed from 2013 President Xi Jinping brought a new emphasis to China, not as a follower in international climate affairs as it had been, but as a '*yinlingzhe*' (a leading state) which is prepared to act in several leadership senses, as a 'pioneer', 'vanguard' and 'leading actor' (Yang, 2022, p. 362). Hurri (2023b) observes that international climate negotiations and China's increasingly significant role in them also presented a golden opportunity for China to increase both its hard and soft power, and support from the Global South, and to show its potential as a global leader. Leadership may also be attributed (Savage, 2019), as it was, for example, to China by significant numbers of the participants at three consecutive Conference of Parties (COP) meetings – in Poznań 2008, Copenhagen 2009 and Cancún 2010, noting, however, the fractured leadership field between the EU, the United States and China (Karlsson et al., 2012). Recognition of the significance of China's role in international climate negotiations pre-dated its 2014 acceptance of the global norm that significant GHG emitters, whether developed or developing countries, must reduce their emissions. As we have seen, however, China's emission reduction is tracking well below its targets, and its 'laggardly' efforts to wean the country off coal as it develops are undermining its leading role in boosting renewable energy (CAT, 2024; CCPI, 2024).

Structural, Entrepreneurial, Cognitive or Exemplary Leadership?

In Chapter 8, we will address the complexities of climate leadership theory and its relevance to our interest in regional climate leadership, but it is worth considering here Li's (2021) observations of China's climate leadership through the lenses of four differing types of leadership – *structural, entrepreneurial, cognitive* and *exemplary* (Wurzel et al., 2021). *Structural leadership* is something of a misnomer for referring to economic power, on the one hand, and the ability on the other to leverage that power into transformative climate change action, and of course, as Li (2021, p. 41) observes, China has considerable structural power. We have seen that until recent crises it has also been the world's fastest growing country and is now the world's second largest economy. However, having structural power does not guarantee structural leadership, particularly not when China is still developing as it attempts to decarbonise and is subject to economic and capacity setbacks. *Entrepreneurial leadership* is the skill of brokering agreement to act on climate change, not only, as the literature suggests, on the international stage where China has been a major player and prominent since 2014, but also at the regional, sub-regional and bilateral levels. Li (2021, p. 41) observes that China's international diplomacy has become increasingly prominent and has included striking bilateral agreements with major emitters, as well as with the developing country coalition, and, as we have seen above, actively engaging with EA&P countries in pursuit of climate action (Table 4.3).

China has assumed *cognitive leadership*, which entails leading by promoting ideas that are transformative but that may take time to be realised. It forcefully argued for a recognition of the common but differentiated responsibility to act on climate change between developed and developing nations. It adopted the pursuit of an ecological civilisation, by which it means the 'embedding of a green agenda' into planning and policy at all levels of interaction from local to global (Li, 2021, p. 40). And its President pledged to strictly control new coal-fired power generation, to promote green technology and renewable energy, and to cease building coal-fired power projects abroad – which is proving elusive in the current economic and geopolitical climate, but which has not been abandoned as a priority. China has also demonstrated *exemplary leadership* which refers to leading by example, whether intentionally or not, notably in China's case by sharing 'its renewable energy best practices and project experience with other developing countries while offering financial aid, technology transfer and personnel training' (Li, 2021, p. 27; Lewis, 2023). As we have seen, China has the world's largest capacity in renewable energy, and its ability to roll it out via its Belt and Road Initiative, its South-South Cooperation projects and its infrastructure funding will likely be a climate change tipping point. However, China's support has been fraught in terms of environmental, geopolitical and security factors, and in terms of its partnering process, sustainable lending and infrastructure practice, all of which must improve if emissions are to be impacted across the globe (Betancourt, 2024).

Conclusions

China has clearly emerged as an ambitious climate leader, helping to shape international efforts to reduce emissions and to avert dangerous climate change; however, the credibility and effectiveness of its efforts remain to be proven. Had it not unequivocally committed to acting in concert with the global community and held fast when the United States vacated the international stage, then the *Paris Agreement* may not have been struck and it may have foundered shortly afterwards. For a range of reasons, in the wake of a series of recent crises, and amidst heightened regional, geopolitical tensions, China is, however, currently failing to reach its near-term climate policy targets, with its efforts being greatly undermined by its burgeoning reliance upon coal-fired power. It has been experimenting with emission trading schemes since 2013 and adopted a nationwide ETS in 2021 (Chapter 8) but this has not proved effective emissions reducing instrumentation at a whole of economy level thus far. On the other hand, China is an undisputed world leader in renewable energy, far outpacing the EU and the United States, and easily eclipsing efforts by the rest of the world and is committed to boosting global renewable energy through its own financing and distribution networks. However, both CCPI and CAT rank China's current efforts as highly insufficient.

We have seen that China has structural leadership by virtue of the size of its economy, if not its capacity, given that it is still a vast, developing nation (Table 5.1), with endemic, multi-levelled implementation challenges (Schreurs, 2017; Engels, 2018). It has nevertheless embraced entrepreneurial leadership in international negotiations and is acknowledged as a country that is crucial to securing the goals of the *Paris Agreement.* China has shown cognitive leadership for promoting significant ideas such as common but differentiated responsibility, ecological citizenship, ending reliance on coal domestically and ending support for it abroad. In terms of leadership theory, China has shown exemplary leadership in many of its climate policy efforts, not least its phenomenal roll-out of renewable energy, but also its integration of climate policy with broader state planning. For various geopolitical and environmental reasons, China has affirmatively assumed a leadership role regionally on climate change action, in terms of cooperation and negotiation with, and direct support of, EA&P countries. It remains at the forefront of efforts to assist developing countries in the region, recognising their need for climate finance, capacity building, and technology transfer via its funding initiatives, although there is some concern about the credibility and effectiveness of its efforts to do so.

8 Conclusions

Regional Climate Leadership in EAP?

Introduction

This book has detailed the extreme vulnerability of the East Asia and Pacific region (EA&P) to climate change impacts with its overpopulation, extreme poverty, high baseline levels of flooding, drought and extreme temperatures, combined with the emissions burden of at times rapid economic development. It has noted the lack of capacity of lesser-developed and developing EA&P countries to respond to climate change impacts given the sheer scale of risk, the contestability of resilience building as a protective action and the inadequate financial support from developed countries. This is despite the internationally accepted principle of differentiation that would see developed countries assuming the financial and policy burden of responding to climate change, given their historical responsibility for the emissions that have caused the problem. Lesser-developed and developing countries have now been drawn into global emissions reduction efforts under the *Paris Agreement*, but, in the EA&P, all country efforts to track towards the *Paris* 2030/2050 emission reduction goals have been inadequate. It is now climate code red for humanity, according to António Guterres, the Secretary General of the United Nations, with, for example, brutal heatwaves across the Asia Pacific as it warms faster than the global average, and as the 'era of global boiling has arrived' (UN, 2023b).

It is in the context of the 'ambitious, urgent, and decisive action by nation states' that is needed to deeply cut emissions if global temperature rises are to be limited to 1.5 °C that is now required (UN, 2021) that we propose a role for regional climate leadership action. A regional leader would act decisively in both domestic and neighbourly contexts, and more broadly within the region to synergistically build its capacity to transition to lower carbon economies more urgently. The criteria by which to measure such efforts would include policy credibility and effectiveness, responsible and coherent domestic and international actions, and actions that deliberately build regional capacity with recognition of differentiated country responsibility. We have noted the complex, fractured institutional context of EA&P, but also the key roles that ASEAN (2023c) and PIF[1] (Pacific Community, 2022), or member countries within these organisations, could play in accelerating decarbonisation and

DOI: 10.4324/9781003170389-8

climate action, as regional climate leadership bodies or nations with significant capacity, although our focus has been upon nation state action. We have also noted Dent's observations that despite deepening regionalism within East Asia, regional community building is weak, and that regional leaders could counter this by acting as champions in solving collective problems like climate change (2008, p. 22).

Theories of Climate Leadership

Regional climate leadership here differs from the climate leadership literature for its focus upon accelerated climate action in regionally specific circumstances, and therefore not just upon leading domestic actions but upon those that synergistically boost regional efforts and build capacity. Our evaluative criteria emphasise that climate leading actions must be credible, effective, responsible, not undermined, and inclusive of support for developing nation efforts. We have focused upon the nation states within EA&P with the greatest capacity to act, namely developed nations, as well as a selection of lesser-developed nations with a better capacity to act than developing nations within the region, rather than upon nations with the least capacity. It matters little which nations we chose, beyond developed nations, in the sense that none of the countries we looked at proved to be regional climate leaders, with our evaluative criteria then offering a set of normative standards for climate ambition and for measuring more authentic future efforts. This exercise has raised the question of the meaning of climate leadership. Invariably, it is reputational because leaders stand out; however, it may be aspirational or intentional, inspiring others to act, and it may also be anecdotal where word circulates of successful efforts which are then transferred or copied, with credit possibly attributed.

Our climate leadership focus has been upon an integrated criteria-based analysis of domestic and regional climate policy efforts to meet the goals of the *Paris Agreement*; however, climate leadership in general is often described as outperforming the efforts of other countries. Wurzel et al. (2021, p. 1) describe environmental leaders, and pioneers, 'as pivotal actors for solving or at least mitigating environmental problems at the domestic and international governance levels'. They offer examples of best practices for followers to emulate, in multilevel and polycentric climate governance structures. Leaders will seek followers, pioneers will not, and followers presumably will not initiate but instead copy the actions of others. However, the study of climate leadership did not originate in concerns with nation state emissions reduction practices but rather in the concern to typologise behaviour in international climate regime formation processes. Positive intellectual leadership was attributed to scientists, for example, for their agenda setting as policy entrepreneurs, whilst key individuals were seen as instrumental leaders negotiating to progress climate regimes (Andresen & Agrawala, 2002). Leadership is also considered an exercise of structural power in climate regime negotiations, by the European Union initially, sporadically by the United States, more recently by China, by the G-77 group of developing countries, and by the Small Island Developing States (Hurri, 2020).

Our focus is upon nation states, given the context of the international agreements to reduce greenhouse gas (GHG) emissions committed to by nation states, and the scientifically informed target setting, reporting, monitoring and improvement processes that are binding upon nation states. Studies of climate leadership typically identify exemplary actions from nation state *leaders* that inspire others to act, from *pioneers* that are at the forefront of the action, or from *followers* that have been motivated to replicate the actions of others. However, they do not extend to questioning the credibility of climate action, the effectiveness of the policy instrumentation adopted, the potential for leading actions to be undermined, or the need to build capacity in the region. The focus is often upon developed nations and their individual target setting and policy innovations. However, climate leadership is not notable on any count in developed nations, despite these nations having the greatest capacity to act. For various reasons, including low per capita emissions, adoption of renewable energy, and policy tracking that aligns with *Paris* targets, the Philippines, India and Nigeria are currently in the top 20 countries on the Climate Change Performance Index (CCPI, 2024). Similarly, Climate Action Tracker ranks Bhutan, Costa Rica, Ethiopia and Kenya, above any developed nation, except for Norway,[2] in terms of climate policies and actions, target tracking and fair share efforts (CAT, 2024).

It is unclear that climate leadership can be precisely calibrated theoretically and therefore measured. For example, is the structural leadership typical of developed nations a useful concept for identifying a climate policy leading nation, given that those countries with structural power, such as the United States, and those rapidly assuming it are very often not climate leaders (Tobin et al., 2021)? Countries such as China, Viet Nam and Costa Rica by contrast show 'entrepreneurial climate leadership, especially at the subnational governance level', whilst Brazil, Indonesia and India show cognitive leadership by championing climate ideas and principles (Tobin et al., 2021, p. 265). However, it is also easy to imagine developed nations exhibiting such leadership more broadly and much more deeply. Multi-case study analysis and longitudinal studies are found by these authors to be necessary to progress and clarify the utility of climate leadership scholarship, and to explore its contradictory findings and caveats that these findings place upon our understanding of leadership typologies. Leadership is not static, for example, but may be 'erratic' and 'contradictory'. Too many climate leaders may lead to 'a fragmented and contradictory approach to global climate governance'. And the study of laggards is underrated given their ability to slow action and reduce global ambition (Tobin et al., 2021, pp. 265–70).

We add to the theory by suggesting that regional leadership has a role to play for its embrace of leading climate actions with both domestic and regional or neighbourly intent, and that outperforms other countries, in a polycentric context and in addressing regional concerns. Given the EA&P's

fractured institutional context, we are not looking for ASEAN or PIF to lead climate policy efforts, although this is desirable and there is scope for it, but rather we are looking for capacity building leadership by nation states within the region. Our normative criteria for measuring regional climate leadership, introduced in Chapter 3, therefore include domestic climate policy credibility and effectiveness, but also responsible and coherent national actions within the region, and action to build regional capacity that recognises common but differentiated country responsibility (Table 3.1). Our approach is policy and outcome focused, seeking to draw lessons, to promote capacity and to suggest solutions, as much as to identify and contrast climate policy performance for a selection of developed, lesser-developed and developing nations in EA&P. Our view is that climate leadership is both a theoretical construct and an applied policy effort and it would do well to acknowledge the regional connectivity, potential connectivity and collaboration between nations to boost mitigation, adaptation and resilience building efforts.

What Is Regional Climate Leadership?

We have an atypical focus on climate leadership within a regional context. In earlier chapters, we established that the EA&P region is the most disaster-stricken region in the world, the most impacted by climate change, by intensifying storms, drought, heavy precipitation, heatwaves, cyclones and the economic and population displacement costs that these trigger (IPCC, 2022b; World Bank, 2022a; WMO, 2023). The United Nations Economic and Social Commission for Asia and the Pacific (UNESCAP) has identified sub-regional 'riskscapes' within the broader Asia Pacific, based upon the IPCC's Sixth Assessment Report, and identified the urgent need for resilience policy pathways. Alarmingly, it finds that climate action has regressed in the region as it faces development challenges in the wake of the COVID-19 pandemic, and recommends that new technologies, climate financing, and regional and sub-regional cooperation are needed (UNESCAP, 2022). The region's decarbonising was tracking in 2022 at only 2.8% compared to the 17.2% that is required to reach net zero by 2050 and needs to accelerate by a factor of at least ten to avert locking in dangerous climate change (World Bank, 2021; PwC, 2023a). However, the region is vulnerable in general, not just to climate change, but because many of its lesser-developed and developing countries do not have the capacity, technology, nor resources to mitigate, adapt to, or, in some cases, to survive the impacts of climate change (IPCC, 2014; WMO, 2023).

The climate change capacity deficit in Asia Pacific is a concern for governance bodies like UNESCAP, the United Nations Development Programme, banks, investment bodies like the World Bank, the Asian Development Bank, and the International Monetary Fund, and insurance companies, international NGOs, research institutions and networks. Despite this, climate leadership theory has been applied to China, recognising the immense challenges it faces as the world's largest GHG emitter, but typically not to the Asia Pacific. Wurzel et al. (2021) do include an analysis of several Asia-Pacific countries in their nation state case study collection. China is found to be a global climate power, with bouts of structural and entrepreneurial leadership, and active innovation at the city level, but has problems in terms of the limits of its top-down climate policy implementation (pp. 41–2). Viet Nam's strong governance and *Green Growth Strategy* are found to underpin its exemplary, entrepreneurial domestic climate leadership and pioneering of solar and wind energy (p. 79). New Zealand prides itself on moral leadership but is found to be a rhetorical climate leader at best because despite 'having passed the world's second-ever Zero Carbon Act', it lacks the policies and will to implement it (pp. 93–5). And Indonesia, one of the world's highest GHG emitters, is neither a climate leader nor an innovator except for sporadic forest governance reform inspired by a REDD (reducing emissions from deforestation and forest degradation) agreement with Norway (pp. 109–11).

Our approach puts individual climate leadership efforts into the context of the regional impacts of climate change, the capacity to respond to it, the challenge of resilience building and of transitioning to lower carbon economies that the aims of the *Paris Agreement* require. It acknowledges that differing countries, in differing geographic locations, with differing levels of development, and differing degrees of reliance upon fossil fuels, will face differing challenges in terms of exhibiting climate leadership. It singles out the extra efforts required from China, from developed countries, from those with high per capita emissions (Australia, South Korea and Japan), and from the key coal producers, consumers, importers and exporters (China, Australia, Indonesia, Viet Nam, the Philippines, Japan, Malaysia and Thailand). It highlights the crucial role that climate finance from developed nations in the region, but also drawn upon more broadly in a polycentric fashion, should play in building resilient lower carbon economies and in pursuing the goals of the *Paris Agreement* by lesser-developed and developing nations. And it highlights the need to phase out coal and to implement energy transition pathways (Table 2.2) most urgently by coal-centric nations such as China, India, Indonesia, Viet Nam and Japan, which could otherwise push the global goals of carbon neutrality beyond reach.

East Asia and the Pacific is a vast region, with no unified, structured, rule-based governance by which to systematically initiate, advance, monitor or improve climate change policy efforts and outcomes. East Asia alone is a region of many regions both formal and informal as Dent (2008) and Beeson (2009) have observed, with regionalism more advanced in East Asia than in the Pacific. The many sets of overlapping climate policy pathways are an ongoing work in progress. These are nevertheless crucial settings for our consideration of leading EA&P nation state action. We have identified that both ASEAN and PIF offer supportive contexts for bilateral, polycentric, and multilevel climate policy, emission reduction and adaptation efforts, with their vision statements, strategies, plans and implementation pathways (Table 2.1). Further climate policy leverage for capacity building is possible through bilateral and broader based partnerships with developed countries within EA&P, particularly those with neighbourly relations and established aid and finance programmes. ASEAN and PIF are facilitators rather than leaders of regional climate action; however, despite lacking the institutional formality of the European Union, both are well able to champion the climate change challenges, needs and vulnerabilities of their region and its member states within international negotiations and contexts, and to guide the transition towards regional goals aligned with the *Paris Agreement*.

The Asia Pacific is a contingent notion. We adopted the Bureau of East Asian and Pacific Affairs' definition of East Asia and the Pacific because of its inclusion of countries within neighbourly contexts of our respective home countries of Australia and Japan. This definition also affords an opportunity, previously identified (Crowley & Nakamura, 2017), for examining developed countries as catalysts and supporters of lesser-developed and developing countries in terms of achieving accelerating climate action within the EA&P. We then chose Australia, China, Indonesia, Japan, Malaysia, New Zealand, South Korea, ROC Taiwan (Taiwan) and Thailand as those countries that appear most economically capable, in terms of their GDP, of reducing their GHG emissions, and of assisting others in the region to do likewise (Table 1.2). This selection of developed, lesser-developed countries and developing countries then became the focus of a more detailed analysis against our regional leadership criteria in Chapter 4. These countries also tended to have higher per capita emissions in the EA&P, the highest accelerating emissions post 1990, and a reliance upon fossil fuels, although the entire region is endemically reliant upon fossil fuels. This is so even of the Pacific Island countries. But their small, geographically isolated communities and economies generate negligible emissions. They are the victims of the accelerating impacts of climate change, and not expected to lead, but we nevertheless look for and find evidence of climate leadership in their demeanour and action.

Leadership by EA&P Developed Countries?

Leadership on climate change in EA&P is expected from countries such as Japan, South Korea, Australia, and to a lesser degree New Zealand, which share the highest GDP in the region (Table 1.2) and high levels of adaptive capacity and climate governance (Table 5.1). If the EA&P is to meet the goals of the *Paris Agreement*, then these countries must lead domestically and in terms of regional capacity building. However, the Climate Action Tracker rates the policies, actions and targets of these countries as insufficient and their climate finance as insufficient to highly insufficient, whilst the Climate Change Performance Index 2024 ranks their performance, respectively, at 58th, 64th, 50th and 34th out of 67 countries. Whilst their collective contribution to global GHG emissions is currently low, their various offsetting schemes, contributions to emissions beyond their own countries, and fossil fuel dependencies are grounds enough for these more capable countries to act more decisively. Of these countries, Japan and Australia have the greatest need to reduce their fossil fuel dependency by adopting a fossil fuel phase out policy, by ceasing subsidies of and approvals for coal mines, and by revoking policies, including offsetting policies, that prolong their reliance upon coal (Table 2.2).

We noted in Chapter 4 the CAT findings that these countries lack credible, effective emission reducing targets, policies and instrumentation and that, if all countries replicated their performance, then global warming would reach 3 °C to 4 °C by 2050 (CAT, 2024). Their policies lack credibility where they rely upon domestic and international offsets rather than transformative practices to achieve low carbon economies, to reduce emissions and to transition away from fossil fuels. Table 4.2 shows the lack of ambition in renewable energy goal setting by Japan, South Korea and New Zealand, and despite Australia's ambitious 82% renewable energy by 2030 goal, progress there is off track in terms of implementation and a failure to phase out coal and gas (CAT, 2024). In terms of transition planning to net zero emissions by 2050, only Australia has committed to an explicit transition plan, the *National Energy Transformation Partnership*, and the complementary *Powering Australia* policy. Neither is there compelling leadership from these countries in terms of accepting regional responsibility for urgent mitigation and transitioning away from fossil fuels for domestic consumption and/or for export revenues. These countries have accepted differentiated responsibility as signatories to UNFCCC agreements and have established trade and aid relationships regionally but thus far without offering regional assistance sufficiently aligned with the aims of the *Paris Agreement*.

In Chapter 5, we discussed the role that developed countries could play in climate preparedness, adaptation and policy resilience building with EA&P, and in Chapter 6 the role that carbon trading could play in achieving bona fide emissions reductions. Challenges requiring regional leadership from developed EA&P countries include the immense and intensifying vulnerability of the region, and the need for building resilience, for dedicating the required resources to doing so and for delivering adequate climate finance. Equally challenging is the emerging carbon market within EA&P, and the need to adopt best practice carbon pricing that triggers significant change in the behaviour of the emitters of fossil fuels. In Chapter 5 we saw that of the US$362 billion per annum in climate finance that is required by the Asia-Pacific region to respond to climate impacts and risks, only US$100 billion per annum is being sought by the *Paris Agreement*, with only a fraction of this being delivered. CAT (2024) finds that climate aid from Japan and New Zealand is highly insufficient, from Australia is critically insufficient, and, whilst South Korea is not assessed, it finds that it has contributed US$600 million to the UN's Green Climate Fund (Table 5.2). In Chapter 6, we saw that there are no regional leaders in carbon trading in EA&P's developed countries. Japan, South Korea, Australia and New Zealand have adopted carbon trading schemes only to fail to develop them, to ensure their efficacy or to secure their longevity.

Each of these developed countries considers themselves leaders within their own immediate region and, to some extent, across the EA&P and has a high degree of global and regional connectivity in prestigious bodies as a measure of their structural influence. Japan is, for example, a G7 country, and Japan, South Korea and Australia are each a member of both the G20 and the G7 Climate Club, which was recently established to advance climate change mitigation policies, to advance industrial decarbonisation, and to boost international climate cooperation and partnerships (Crippa et al., 2023). In Chapter 2, we saw that there are multiple regional pathways for advancing climate policy efforts and for achieving enhanced outcomes (Table 2.1). Japan, South Korea, Australia and New Zealand are members of ASEAN+6, the ASEAN Regional Forum, the East Asia Summit and the Asia-Pacific Economic Cooperation council. Whilst Australia and New Zealand are long-standing members of the Pacific Island Forum and the SPC Pacific Community, Japan, South Korea, Australia and New Zealand are all members of the Pacific Economic Cooperation Council (Table 2.1). Climate change mitigation, adaptation, resilience and capacity building are concerns of these bodies, but further research is needed to consider why the 'climate club' opportunities they offer are not being fully and urgently exploited by developed countries as a means of accelerating the implementation of the *Paris Agreement.*

In Chapter 1, we saw that developed countries are amongst the highest per capita emitters of GHG emissions in the EA&P, but also the most *economically* capable of leading regional climate change efforts, although their structural reliance upon coal and gas is a significant economic obstacle. In Chapter 4, we saw that they are committed to the *Paris Agreement* and its 1.5 °C goal, with only Australia committed to a 2 °C limit, but that their policy efforts lack credibility and effectiveness and that they are not inspiring or supporting lesser-developed or developing nations within the region to act. Their emission reduction and renewable energy targets are incompatible with the *Paris* goals, and their reliance upon fossil fuels is structured into their economic and development planning. They have no viable emissions reducing policy instruments, with carbon pricing offering unrealised potential, they are overly reliant upon offsetting rather than reducing emissions, and they have built offsetting into their carbon accounting processes (Table 4.1). None of these countries is a regional climate leader, and whilst there is scope for them to elevate their domestic and regional climate change efforts and to assume leadership, they would clearly do so only if it were in their own national interests.

Leadership by Lesser-developed and Developing Countries?

Whilst the *Kyoto Protocol* had embraced prior action by developed countries (Table 3.1), the failure of these countries to effect sufficient emissions reductions has seen the less affluent, lesser-developed and developing countries drawn into the global emissions reduction effort under the *Paris Agreement.* Despite this, lesser-developed and developing countries are considerably less capable of climate change mitigation, adaptation, resilience and capacity building, and we have argued that a regional climate leader would acknowledge this and help build their efforts. We saw in Chapter 2 that the EA&P remains endemically reliant upon coal, not only with 85% of its energy consumption sourced from fossil fuels but also with coal generation in the region expanding and not expected to phase out before 2060 (World Bank, 2023a; IMF, 2021a). But how is Indonesia, for example, with a GPD that is only one quarter of Japan, South Korea, Australia and New Zealand's, to extricate itself from its role as the world's number one coal exporting nation, and from its domestic consumption which has risen by 10% per annum (Table 1.2)? And why aren't developed countries within the EA&P proactively engaging with China, Taiwan, Viet Nam, the Philippines and Thailand, which are nearly all top ten coal importing nations (Table 2.2), offering partnerships to help them phase out their reliance upon coal and to transition to renewable energy and sustainable, net zero economies?

In Chapter 5, we saw that EA&P is already the most disaster-stricken region in the world and that climate change impacts are outstripping its adaptation and resilience building efforts whilst increasing natural disasters and their intensity and driving up involuntary population migration (IPCC 2022b; WMO, 2023). The global costs of extreme events are estimated at US$160 billion to $340 billion per annum by 2030, and US$315 billion to US$565 billion per annum by 2050, with residual risks and unavoidable loss and damage likely to double these amounts (UNEP, 2022). Table 5.1 shows how a low-income EA&P country like the Philippines, which is now the country in the region that is most affected by natural disasters year after year (Eckstein et al., 2021), has a heightened climate risk, but a lowly ranked climate readiness, adaptive capacity, and governance. The lower income Pacific nations of Micronesia, Tonga, and the Solomon Islands are the most vulnerable to climate change, and the least ready to meet its challenges, with the lowest adaptive capacities and generally with less governance capacity (Table 5.1; ND-GAIN, 2022). Equally in East Asia, the developing, lower income economies such as Thailand, the Philippines, Malaysia and Indonesia have the lowest adaptive capacity, the greatest vulnerability to extreme weather, and will experience the greatest risk and highest impact on their GDP (Higginbotham, 2021; Table 5.1).

Against this challenging backdrop, we found in Chapter 5 that lesser-developed and developing countries in the Asia Pacific are receiving one tenth of the US$362 billion per annum that is needed to meet the *Paris Agreement* goals (UNESCAP, 2023c, p. 51). To achieve the region's 2030 *Paris* goals, climate finance would need to be boosted by c.590% (CPI, 2021). In these straightened and strained circumstances, and despite no expectations of climate leadership from lesser-developed and developing countries, we find in Chapter 4, for example, that Malaysia has voluntarily established a net zero by 2050 target, a national energy transition roadmap, a carbon exchange, and is exploring establishing carbon pricing. Taiwan lacks nation state status and so only has UNFCCC observer status, but it has exhibited significant 'intentional' leadership tendencies including at annual COP meetings, and in voluntarily adopting a legally binding net zero by 2050 target, establishing a carbon exchange and engaging in bilateral and regional climate dialogue. Indonesia is committed to meeting its conditional (with international assistance) and unconditional (without international assistance) targets, Thailand has strengthened its ambitions and launched a voluntary carbon credit exchange, and Viet Nam has legislated its net zero target and its carbon market. Nevertheless, for reasons of fossil fuel dependency, reliance upon offsetting emissions, questionable policy instrumentation, implementation and domestic political issues, and inadequate target setting, such efforts, and those of China, are ranked by CAT (2024) as highly to critically insufficient.

Of all the countries we have considered, China is the most likely emergent regional climate leader, although with important caveats, including its current climate policy slowdown, its prolonged reliance upon fossil fuels, and without consideration of its domestic political context. Climate Action Tracker (CAT, 2024) rates China's climate policy performance as highly insufficient for its lack of ambitious target setting, long-term strategy with climate policies and measures, and lack of plans to phase out fossil fuel, whilst the Climate Change Performance Index ranks China as 51st of 67 countries (CCPI, 2024). It is still a developing country with a low carbon footprint, but China's emissions have burgeoned by 203% since 1990, and it is now the world's largest emitter of GHG emissions (Table 1.2). However, it aims to peak emissions by 2030, and to reach carbon neutrality before 2060, and there are currently nine subnational ETSs and one nationwide ETS in the country with plans for expansion. China has released *An Energy Sector Roadmap to Carbon Neutrality in China* including an *Accelerated Transition Scenario* that aligns carbon neutrality with its broader development goals (increasing prosperity and innovation-driven growth). It is highly engaged in international, regional and bilateral climate policy dialogues and in facilitating global renewable energy uptake and is a significant climate change financier around the world as well as in the EA&P region through its banks, its Belt and Road Initiative and its South-South Cooperation Fund (see Chapters 4 and 7).

We have found clear and at times unexpected grounds for including lesser-developed and developing nations in comparative and case study analysis of climate leadership, or in our case of regional climate leadership. Beyond the most obvious example of China, which we found to be a significant, but nevertheless Janus-faced leader with currently faltering efforts but immense potential future impact, there is evidence of emergent leadership beyond developed countries. The most understated regional climate leadership is from the Pacific Island nations, with their decades of agenda-setting international level lobbying for the rights of small island developing states, including their declaration of a climate emergency and lobbying for a global fossil fuels non-proliferation treaty (FFNPTI, 2024b). We also found, in Chapter 4, that Fiji, which was instrumental with Vanuatu and Tuvalu in spearheading this treaty and is not expected to act significantly on climate change, has exceeded the climate policy ambitions, policy efforts and expectations of a small island developing nation. Indeed, we found that many lesser-developed and developing EA&P countries are voluntarily adopting *Paris* aligned targets and are voluntarily reporting against them, although many of their efforts are ranked highly or critically insufficient by Climate Action Tracker. Taiwan is a notable and novel example, given its lack of country status, for its international, regional and domestic efforts, but there are signs of leading developing nation climate actions across the entire EA&P region.

Regional Leadership in East Asia and the Pacific?

In China's case, and across the EA&P countries we have analysed, we have found climate leadership to be a contradictory notion. It is not the systematic exercise envisaged by our criteria that leading regional climate policy actions are credible, effective, responsible, not undermined, and inclusive of support for developing nation efforts. The credibility gap is notable, where a country has ambitious target setting and policy aspirations, but also undermining actions, in prolonging fossil fuels, for example, or choosing ineffective policy instrumentation (Chapter 6). Except for the efforts of China, Taiwan and the Pacific Island Forum, we saw no comprehensive leading climate policy efforts to promote mitigation, adaptation and resilience actions beyond the nation state and within the region. Future research beyond our comparative analysis with detailed state by state case study analysis, such as our consideration of China, would paint a fuller picture of nation state efforts. We understand that EA&P countries are poorly ranked by Climate Action Tracker and the Climate Change Performance Index which are produced by consortia of leading climate change research institutes, think tanks and international non-government climate organisations. However, our analysis is focused upon not only performance but also upon the varying capacities of developed, lesser-developed and developing EA&P countries to meet the goals of the *Paris Agreement.*

The context for this study is that parties to the *Paris Agreement* are not on track to meet its 2030 or 2050 goals and that, given the accelerating impacts of climate change, a rapid shift is needed to accelerate the transition to a low carbon economy and a less carbon-endangered world. The IPCC has observed that we have the tools to do this, in acknowledging the proliferation of policies and laws that have enhanced energy efficiency, reduced the rates of deforestation and accelerated the deployment of renewable energy. However, the countries of EA&P account for nearly half of global emissions, with China accounting for the bulk of those, so if these countries don't respond adequately and swiftly, then the goals of the *Paris Agreement* will be beyond reach. For the region to decarbonise by a factor of at least ten times the current effort, which the World Bank (2021) estimates is needed, credible and effective, EA&P efforts must be connected in a polycentric fashion, cooperative beyond the nation state, and inclusive of tangible financial and technological support for lesser-developed and developing nation efforts. Regional leadership in driving this change would not substitute for domestic action but could have a synergistic impact whereby initiatives across the region are stimulated more immediately and effective policy is readily transferred.

We have found, of course, that climate change action imposes a universally disrupting cost, particularly disrupting of fossil fuel dependency, that is beyond the remit of the lesser-developed and developing nations within EA&P, and that nation state interests predominate. This can generate perverse policy impacts, for example, in the experimentation with and the implementation of carbon pricing within the region where this is relied upon as a simplistic climate policy panacea. There is polycentric support for best practice carbon pricing schemes from international governance bodies, like the United Nations, the OECD, and the international banking sector, including the World Bank, IMF and the Asian Development Bank, as well as research institutes and non-government bodies. Despite this, nation state interests, and not policy logic, influence the adoption of carbon pricing. There have been pioneering efforts from Japan for setting the carbon pricing agenda, New Zealand for following, Australia for pioneering a comprehensive scheme, China for its novel regional experimentation, and South Korea, Thailand and Indonesia for following. But there have been setbacks and disappointments. Schemes have been poorly implemented, underdeveloped, undermined, abandoned or unsupported by complementary policies, so that the price triggering impact upon the behaviour of emitters is lost, and the focus shifts to trading and revenue raising. The carbon pricing conundrum raises the issue of policy displacement, for example, where fossil fuel longevity is built into climate policies and instrumentation, hampering the transition to renewables, and again placing the *Paris Agreement* goals beyond reach.

Our study of regional climate leadership has thrown up meagre results in terms of recognisable leadership actions, although there are sporadic leading, pioneering and transferred efforts, and China is indeed clearly a climate superpower and emergent global and regional leader. It has, however, illustrated the climate change challenges that EA&P is facing, the regional and domestic contexts for action, and the efforts that are required to exit fossil fuels, to finance lesser-developed and developing nation efforts, and to secure lower carbon economies. It is also worth noting that the supportive context that ASEAN and PIF each offer for nation state and collaborative action is very different in the emissions intensive East Asia, to the relatively coal free Pacific where climate change is an existential threat to daily life. If climate clubs offer a swifter path to accelerated action, then there are potentially crucial roles for ASEAN and PIF to play in more actively advocating for and supporting regional climate leadership by their members and with the support of partnering developed states. Potentially they could set regional emission reduction and renewable energy targets with differentiated expectations of the most and the least capable countries within EA&P and establish climate clubs for those countries that achieve against these targets. Currently, emission reductions targets within the region are collectively no match for those within the EU, for example, and the articulation of EA&P regional or sub-regional goals or targets may inspire heightened ambitions.

Conclusions

The EA&P is not replete with developed countries. It has some way to go to meet the challenges of 'economic growth, poverty reduction, food security, public health, and jobs' that institutional partners such as the World Bank (2024a) are actively assisting with. The region must also rapidly transition to renewable energy and low carbon economies if the *Paris Agreement* is to be realised; however, not even the developed countries in EA&P are managing to do this. We have argued that the problem of climate change is too acute to be waiting for unilateral leadership to be demonstrated at the nation state level, because there is less dynamic for replicating unilateral efforts than there is for replicating efforts in polycentric circumstances. Instead, we have emphasised the role that nation states can play by following their domestic efforts, successes and innovations, with broader, regional, bilateral, multilateral and potentially global efforts that seek connectivity, collaboration and rapid policy transfer. Within EA&P, this is not just an existential environmental challenge, but an existential development challenge, with its emergent economies expected by the international community to by-pass the fossil fuel led growth that established today's developed countries. We have suggested that regional leadership is needed within EA&P, with regional climate leaders acting decisively in domestic and neighbourly contexts, and more broadly within the region, to synergistically build its capacity to transition to lower carbon economies more urgently.

The problem we encountered is that there are no outstanding EA&P leaders demonstrating policy credibility and effectiveness, responsible and coherent national regional actions, nor actions to build regional climate capacity that recognises common but differentiated country responsibility. We found the same 'erratic' and 'contradictory' climate leadership efforts, undermined at times by 'laggardly' efforts, as did Tobin et al. (2021, pp. 265–70), with the same implications that these may lead to 'a fragmented and contradictory approach to global climate governance'. However, we also found that the UNFCCC's agreement, monitoring and reporting processes are invaluable agenda setting tools in terms of staking out global and domestic ambitions, as well as providing the guardrails for the development of increasingly directive accountability processes. However much developed nation states seek to shirk their climate policy responsibilities, we found that these processes translate directly into boosting the efforts of small lesser-developed nations like Taiwan, Viet Nam, Fiji and the relatively tiny Pacific Island countries. We were reluctant to let China distract us from our survey of lesser EA&P nations; however, we found China to be a climate policy monolith with the capacity to entirely transform global climate and energy policy and its impacts. We expected to find developed nations leading in the race away from regional climate disaster;

however, future research may well attribute regional climate leadership to an unexpected pivot by the lesser-developed and developing nations as a means fast-tracking their development trajectories (ASEAN, 2023c).

Notes

1 ASEAN Association of Southeast Asian Nations. PIF Pacific Island Forum (see Chapter 2).
2 Noting that exported emissions from Norwegian oil and gas are not considered in CAT's rating.

References

Abe, T., & Arimura, T.H. (2022). Causal effects of the Tokyo emissions trading scheme on energy consumption and economic performance. *Energy Policy*, 168, 113151.

Adger, W.N. (2000). Social and ecological resilience: Are they related? *Progress in Human Geography*, 24 (3), 347–364.

Ahmed, A.U., Appadurai, A.N., & Neelormi, S. (2019). Status of climate change adaptation in South Asia region. In M. Alam, J. Lee & P. Sawhney (Eds.), *Status of climate change adaptation in Asia and the Pacific.* (pp. 125–152). Berlin Heidelberg: Springer.

Andresen, S., & Agrawala, S. (2002). Leaders, pushers, and laggards in the making of the climate regime. *Global Environmental Change*, 12 (1), 41–51.

Aoki, K. (2022). *Viet Nam announces national strategy to achieve net-zero greenhouse gas emissions by 2050.* August 5. Tokyo: EnvilianceAsia. Retrieved from https://enviliance.com/regions/southeast-asia/vn/report_7617.

APEC Secretariat. (2022). *APEC stocktake of carbon pricing initiatives.* Singapore: APEC. Retrieved from https://www.apec.org/docs/default-source/publications/2022/2/apec-stocktake-of-carbon-pricing-initiatives/222_ec_apec-stocktake-of-carbon-pricing-initiatives.pdf.

Arežina, S. (2019). U.S.-China relations under the Trump administration: Changes and challenges. *China Quarterly of International Strategic Studies*, 5 (3), 289–315.

ASEAN-China. (2021). *ASEAN-China joint statement on enhancing green and sustainable development cooperation.* Singapore: APEC. Retrieved from https://asean.org/wp-content/uploads/2021/10/65.-Final-ASEAN-China-Joint-Statement-on-Green-and-Sustainable-Development-Cooperation.pdf.

Asia Pacific Centre for Security Studies (APSCC) (2023). *Countries of Asia Pacific.* Honolulu: Asia-Pacific Centre for Security Studies.

Asia-Pacific Economic Cooperation (APEC) (2015). *APEC Energy Ministers push for energy resilience.* Singapore: APEC. Retrieved from https://www.apec.org/press/news-releases/2015/1013_statement.

Asia-Pacific Economic Cooperation (APEC) (2023). *Strong, balanced, secure, sustainable, and inclusive growth.* Singapore: APEC. Retrieved from https://aotearoaplanofaction.apec.org/strong-balanced-secure-sustainable-and-inclusive-growth.html.

Asia Society Policy Institute (ASPI) (2023). *Climate action brief: China.* New York: ASPI.

Asian Development Bank (ADB) (2019). *Climate change and disasters in Asia and the Pacific.* Manila: ADB.

Asian Development Bank (ADB) (2023). *Climate finance landscape of Asia and the pacific.* Manila: ADB.

Association of Southeast Asian Nations (ASEAN) (2010). *ASEAN leaders statement on joint response to climate change Ha Noi April 2010.* Jakarta: ASEAN. Retrieved from

https://asean.org/asean-leaders-statement-on-joint-response-to-climate-change-ha-noi-9-april-2010/.
Association of Southeast Asian Nations (ASEAN) (2015). *ASEAN 2025: At a glance.* Jakarta: ASEAN. Retrieved from https://asean.org/asean-2025-at-a-glance/.
Association of Southeast Asian Nations (ASEAN) (2020). *ASEAN State of climate change report.* Jakarta: ASEAN. Retrieved from https://asean.org/book/asean-state-of-climate-change-report/.
Association of Southeast Asian Nations (ASEAN) (2021). *East Asia Summit Leaders' statement on sustainable recovery.* Jakarta: ASEAN. Retrieved from https://asean.org/east-asia-summit-leaders-statement-on-sustainable-recovery/.
Association of Southeast Asian Nations (ASEAN) (2022a). *ASEAN Plus Three cooperation work plan 2023-2027.* Jakarta: ASEAN. Retrieved from https://asean.org/asean-plus-three-cooperation-work-plan-2023-2027/.
Association of Southeast Asian Nations (ASEAN) (2022b). *ASEAN chairman's statement of the 29th ASEAN Regional Forum.* Jakarta: ASEAN. Retrieved from https://asean.org/chairmans-statement-of-the-29th-asean-regional-forum/.
Association of Southeast Asian Nations (ASEAN). (2023a). ASEAN charts course for a sustainable future with ambitious ASEAN Strategy for carbon neutrality. Jakarta: ASEAN. Retrieved from: https://asean.org/asean-charts-course-for-a-sustainable-future-with-ambitious-asean-strategy-for-carbon-neutrality/.
Association of Southeast Asian Nations (ASEAN) (2023b). *ASEAN Regional Forum.* Jakarta: ASEAN. Retrieved from https://aseanregionalforum.asean.org/about-arf/.
Association of Southeast Asian Nations (ASEAN) (2023c). *ASEAN charts course for a sustainable future with ambitious ASEAN Strategy for Carbon Neutrality.* Jakarta: ASEAN. Retrieved from https://asean.org/asean-charts-course-for-a-sustainable-future-with-ambitious-asean-strategy-for-carbon-neutrality/.
Australian National University College of Law (ANUCOL) (2022). *Australia's carbon market a 'fraud on the environment'.* Canberra: ANU College of Law. March 24. Retrieved from https://law.anu.edu.au/news-and-events/news/australias-carbon-market-fraud-environment.
Averchenkova, A., & Bassi, S. (2016). *Beyond the targets: Assessing the political credibility of pledges for the Paris Agreement* (Policy Brief). London: Grantham Research Institute on Climate Change and the Environment and Centre for Climate Change Economics and Policy.
Bahardur, A.V., Ibrahim, M., & Tanner, T. (2010). *The resilience renaissance: Unpacking of resilience for tackling climate change and disasters.* Brighton, UK: Institute of Development Studies, University of Sussex.
Balcazar, C.F., & Kennard, A. (2023). Climate change and political mobilization: Theory and evidence from India. Unpublished paper. September 1. Retrieved from SSRN: https://ssrn.com/abstract=4206967.
Baranzini, A., Jeroen, I., van den Bergh, C.J.M., Carattini, S., Howarth, R.B., Padilla, E., & Roca, J. (2017). Carbon pricing in climate policy: Seven reasons, complementary instruments, and political economy considerations. *WIREs Climate Change*, 8 (4), e462.
Beasley, B. (2022). Climate prepared countries are losing ground, latest ND-GAIN Index shows. University of Notre Dame. October 5. Notre Dame: Global Adaptation Initiative, University of Notre Dame.
Beeson, M. (2009). *Institutions of the Asia-Pacific: ASEAN, APEC and beyond.* New York: Routledge.
Betancourt, M. (2024). Can the Belt and Road go green? *Eos Magazine*, March 7.
Birkland, T. (2010). Federal disaster policy: Learning priorities and prospects for resilience. In L.K. Comfort, A. Boin & C.C. Demchak (Eds.), *Designing resilience: Preparing for extreme events.* (pp. 106–142). Pittsburgh: University of Pittsburgh Press.

Boudreau, C. (2023). Fossil fuel nations are the new nukes: Pacific Island nations at risk from rising sea levels support a non-proliferation treaty for oil and gas. *Business Insider*, March 24.

Brown, H.C.P., Nkem, J., Sonwa, D.J., & Bele, Y. (2010). Institutional adaptive capacity and climate change response in the Congo Basin forests of Cameroon. *Mitigation and Adaptation Strategies for Global Change*, 15, 263–282.

Bureau of East Asian & Pacific Affairs (BEA&PA) (2017). *East Asian and Pacific affairs: Countries and other areas.* Archive. Washington, DC: US Department of State.

Cain, T.N., Kant, R., Ruwet, M., & Byrne, C. (2022). Climate conversations and disconnected discourses: An examination of how Chinese engagement on climate change aligns with Pacific Priorities. Regional Outlook Paper No. 69. Griffith University Asia Institute.

Carbon Brief. (2015). *Explainer: Why 'differentiation' is key to unlocking Paris climate deal.* December 7. London: Carbon Brief. Retrieved from https://www.carbonbrief.org/explainer-why-differentiation-is-key-to-unlocking-paris-climate-deal/.

Carbon Brief. (2021). Which countries are historically responsible for climate change? London: Carbon Brief. October 5. Retrieved from https://www.carbonbrief.org/analysis-which-countries-are-historically-responsible-for-climate-change/.

Carbon Brief. (2022). Guest post: Is China living up to its pledges on climate change and energy transition? London: Carbon Brief. Retrieved from https://www.carbonbrief.org/guest-post-is-china-living-up-to-its-pledges-on-climate-change-and-energy-transition/.

Carbon Brief (2023). *The Carbon Brief profile: China.* London: Carbon Brief.

Carbon Tracker (2021). *Do not revive coal: Planned Asia coal plants a danger to Paris.* London: Carbon Tracker Initiative.

Cassella, C. (2019). There's a climate threat facing Pacific Islands that's more dire than losing land. *Science Alert*, September 19. Retrieved from https://www.sciencealert.com/pacific-islanders-are-in-a-climate-crisis-as-rising-sea-levels-threaten-water.

Cazorla, M., & Toman, M. (2000). *International equity and climate change policy. Climate Issue Brief No. 27.* Washington, DC: Resources for the Future. Retrieved from https://media.rff.org/documents/RFF-CCIB-27.pdf.

Central Intelligence Agency (CIA) (2024). *Country comparisons – Population.* Langley: CIA. Retrieved from https://www.cia.gov/the-world-factbook/field/population/country-comparison/.

Centre for Research on Energy and Clean Air (CREA) (2024). China risks missing multiple climate commitments as coal power approvals continue. Helsinki: CREA. Retrieved from https://energyandcleanair.org/wp/wp-content/uploads/2024/02/CREA_GEM_2023H2-coal-power-briefing_China-missing-climate-commitments.pdf.

Cheeseman, P. (2023). *Weather, climate, and catastrophe report: APAC (Asia Pacific) insights.* Sydney: AON. Retrieved from https://aoninsights.com.au/2023-weather-climate-and-catastrophe-report-apac-insights/.

Climate Action Tracker (CAT) (2024). Countries. New York: Climate Analytics; Cologne/Berlin: NewClimate Institute. Retrieved from https://climateactiontracker.org/countries/.

Climate Analytics (2021). *Coal phase out and energy transition pathways for Asia and the Pacific.* (Policy Brief). Berlin: Climate Analytics.

Climate Analytics (2023). *2030 targets aligned to 1.5C: Evidence from the latest global pathways.* (Policy Brief). Berlin: Climate Analytics.

Climate Change Performance Index (CCPI) (2024). *Climate change performance 2024 – Ranking and results (Policy Brief).* Berlin: Germanwatch; Cologne/Berlin: NewClimate Institute; and Bonn: Climate Action Network International. Retrieved from https://ccpi.org/.

Climate Policy Institute (CPI) (2021). *Global landscape of climate finance.* Prepared by B. Naran, P. Fernandes, R. Padmanabhi, P. Rosane, M. Solomon, S. Stout, C. Strinati, R. Tolentino, E. Wakaba, Y. Zhu, & B. Buchner. San Francisco: CPI.

Commonwealth Scientific and Industrial Research Organization (CSIRO) (2006). Climate change in the Asia/Pacific region. A consultancy report prepared for the Climate Change and Development Roundtable. Canberra: CSIRO.

Crippa, M., Guizzardi, D., Pagani, F., Banja, M., Muntean, M., Schaaf, E., Becker, W., Monforti-Ferrario, F., Quadrelli, R., Risquez Martin, A., Taghavi-Moharamli, P., Köykkä, J., Grassi, G., Rossi, S., Brandao De Melo, J., Oom, D., Branco, A., San-Miguel, J., & Vignati, E. (2023). *GHG emissions of all world countries.* Luxembourg: Publications Office of the European Union, doi:10.2760/953322, JRC134504.

Crowley, K. (2007). Is Australia faking it? The Kyoto Protocol and the Greenhouse Policy Challenge. *Global Environmental Politics*, 7 (4), 118–139.

Crowley, K. (2017). Up and down with climate politics 2013–2016: The repeal of carbon pricing in Australia. *Wiley Interdisciplinary Reviews: Climate Change*, 8 (3), 1–13.

Crowley, K., & Nakamura, A. (2015). Regional climate leadership. *Environmental Management*, 51 (5), 44–48 (*in Japanese*).

Crowley, K., & Nakamura, A. (2017). Defining regional climate leadership: Learning from comparative analysis in the Asia Pacific. *Journal of Comparative Policy Analysis: Research and Practice*, 20 (4), 387–403.

Dabla-Norris, D., Daniel, J., Nozaki, M., Alonso, C., Balasundharam, V., Bellon, M., Chen, C., Corvino, D., & Kilpatrick, J. (2021). Fiscal policies to address climate change in Asia and the Pacific. Research paper 21/07. Washington, DC: International Monetary Fund.

De Bruijne, M., Boin, A., & Van Eeten, M. (2010). Resilience: Exploring the concept and its meanings. In L.K. Comfort, A. Boin & C.C. Demchak (Eds.), *Designing resilience: Preparing for extreme events.* (pp. 13–32). Pittsburgh: University of Pittsburgh Press.

Dent, C.M. (2008). *China, Japan, and regional leadership in East Asia.* Cheltenham, Gloucestershire: Edward Elgar.

Dent, C.M. (2012). Regional leadership in East Asia: Japan and China as contenders. In M. Beeson & R. Stubbs (Eds.), *The Routledge handbook of Asian regionalism.* (pp. 263–274). London: Routledge.

Department of Foreign Affairs and Trade (DFAT) (2022). *East Asia Summit.* Canberra: DFAT.

Department of Foreign Affairs and Trade – Australia (DFAT) (2023a). *Supporting the Indo-Pacific to tackle climate change.* Canberra: DFAT.

Department of Foreign Affairs and Trade – Australia (DFAT) (2023b). *Counting Australia's climate finance.* Canberra: DFAT.

Department of Information Services (DIS) (2023). Reaching net-zero emissions by 2050. Executive Yuan Press Release. March 14. Taipei: DIS, Executive Yuan. Retrieved from https://english.ey.gov.tw/News3/9E5540D592A5FECD/5cf73389-61f9-43dd-80b3-62926546d710.

Deutsch's Institut fur Entwicklungspolitik (DIE) (2022). *NDC explorer.* Bonn: German Institute of Development and Sustainability. Retrieved from https://klimalog.idos-research.de/ndc/#NDCExplorer/worldMap?NewAndUpdatedNDC??income??? catIncome.

Di Sario, F., & Leali, G. (2023). Europe takes climate fight global as carbon border tax goes live. *Politico.* October 1.

Donor Tracker (2023). *Donor Tracker Sector: Climate.* Berlin: SEEK Development. Retrieved from https://donortracker.org/topics/climate.

Duggal, V.K. (2020). Carbon market cooperation to build a low-carbon future. In B. B. Susantono & C.Y. Park (Eds.), *Future of regional cooperation in Asia and the Pacific.* (pp. 497–534). Manila: Asian Development Bank.

E3G. (2023). *China's climate finance to developing world falls short of its own pledges.* London: Third Generation Environmentalism Ltd (E3G). Retrieved from https://www.e3g.org/news/china-climate-finance-developing-world-falls-short-of-its-own-pledges/#.

Eckersley, R. (2012). Does leadership make a difference in international climate politics? A working paper for International Studies Association Annual Conference, San Diego, April 1–4.

Eckersley, R. (2020). Rethinking leadership: Understanding the roles of the US and China in the negotiation of the Paris Agreement. *European Journal of International Relation*, 26 (4), 1178–1202.

Eckstein, D., Kunzel, V., & Schafer, L. (2021). Global Climate Risk Index 2021: Weather-related loss events in 2019 and 2000-2019. Briefing paper. Bonn: German Watch.

Econ Digest (2022). Carbon credit market in Thailand: Opportunities for the business sector. October 11. Bangkok: Kasikorn Research Centre.

Egerough-Adindu, I. (2022). Global climate change governance: The efficacy of multilateral approaches. *Beijing Law Review*, 13, 320–339.

Energy Information Administration (EIA) (2022). *Country information – China.* Washington, DC: EIA. Retrieved from https://www.eia.gov/international/analysis/country/CHN.

Energy Institute (2023). *Statistical review of world energy.* London: Energy Institute.

Engels, A. (2018). Understanding how China is championing climate change mitigation. *Palgrave Communications*, 4 (1), 1–6.

Erbach, G., & Jochheim, U. (2022). *China's climate change policies: State of play ahead of COP27.* Brussels: EU Parliamentary Research Service.

Errendal, S., Ellis, J., & Jeudy-Hugo, S. (2023). *The role of carbon pricing in transforming pathways to reach net zero emissions: Insights from current experiences and potential application to food systems.* Paris: OECD.

European Parliamentary Research Service (EPRS) (2015). Negotiating a new UN climate agreement: Challenges on the road to Paris. Brussels: European Union.

Fejka, S. (2012). Offsetting programs: Struggling to find an equitable solution internationally. *Sustainable Development Law & Policy*, 12 (2), 38–39.

Fiji (2020). Republic of Fiji updated nationally determined contribution – Submitted to the UNFCCC. Suva-Nabua: Regional Pacific NDC Hub.

Fitzgerald, J. (2022). Transitioning from urban climate action to climate equity. *Journal of the American Planning Association*, 88 (4), 508–523.

Fossil Fuel Non-Proliferation Treaty Initiative (FFNPTI) (2024a). Pacific Civil Society launches Naiuli Declaration for a Fossil Fuel Treaty at Global Citizen NOW Action Summit. Press release. March 6. California: Earth Island Institute.

Fossil Fuel Non-Proliferation Treaty Initiative (FFNPTI) (2024b). Why do we need a fossil fuels non-proliferation treaty. Berkeley, California: FFNPTI. Retrieved from https://fossilfueltreaty.org/.

Gallopin, G.C. (2006). Linkages between vulnerability, resilience, and adaptive capacity. *Global Environmental Changes*, 16, 293–303.

Gammon, L. (2022). ASEAN's operation offers a guide to Pacific nations. *Australian Financial Review*, August 21.

Gaspar, K.V. & Yong Rhee, C. (2021). Asia-Pacific, the gigantic domino of climate change. March 25. Washington, DC: International Monetary Fund. Retrieved from https://www.imf.org/en/Blogs/Articles/2021/03/25/asia-pacific-the-gigantic-domino-of-climate-change.

Gerretsen, I. (2023). UN body makes 'breakthrough' on carbon price proposal for shipping. Climate Home News, May 23. Kent: Climate Home News.

Global Energy Monitor (GEM) (2023). Boom and bust coal 2023: Tracking the global coal plant pipeline. [Published in association with Global Energy Monitor, CREA, E3G, Reclaim Finance, Sierra Club, solutions for our climate. With Kiko Network, CAN Europe, Bangladesh Groups, Alliance for Climate Justice & Clean Energy, and Chile Sustentable. San Francisco: GEM.

Green Finance and Development Centre (GFDC) (2023). Countries of the Belt and Road Initiative. Shanghai: Fudan University. Retrieved from https://greenfdc.org/countries-of-the-belt-and-road-initiative-bri/.

Green, F., & Finighan, R. (2012). Laggard to leader: How Australia can lead the world to zero carbon prosperity. Fitzroy, Victoria: Beyond Zero Emissions.

Grojean, W., Resick, C.J., Dickson, M.W., & Smith, D.B. (2004). Leaders, values, and organizational climate: Examining leadership strategies for establishing an organizational climate regarding ethics. *Journal of Business Ethics*, 55 (3), 223–241.

Grossman, D. (2023). South Korea's surprisingly successful China policy. *The Rand Blog*. November 27.

Gunderson, L.H., & Holling, C.S. (2002). *Panarchy: Understanding transformations in human and natural systems*. Washington, DC: Island Press.

Habib, B. (2011). *Climate change and international relations theory: Northeast Asia as a case study*. Paper presented at the World International Studies Committee Third Global International Studies Conference, August 17–20, University of Porto, Portugal.

Hall, D. (2021). Why a carbon price alone won't be enough to drive down New Zealand's emissions. *The Conversation*. June 17.

Harris, B. (2014). Repeating the failures of carbon trading. *Pacific Rim Law & Policy Journal*, 23 (3), 755–793.

Hawkins, A. (2023). China ramps up coal despite carbon neutral pledges. *The Guardian*, April 24.

Hawkins, A. (2024). Growth in CO_2 emissions leaves China likely to miss climate targets. *The Guardian*, February 22.

Haye-ha, L. (2023). Yoon pledges additional US$300 mill to Green Climate Fund at G20 session. *Yonhap News*. September 9. Retrieved from https://en.yna.co.kr/view/AEN20230909001100315.

Higginbotham, R. (2021). Climate change: The real costs to Asia Pacific. Insurance Asia News, April 23. Zürich: Swiss Re Group.

Hovi, J., Sprinz, D.F., & Underdal, A. (2009). Implementing long term climate policy: Time inconsistency, domestic politics, international anarchy. *Global Environmental Politics*, 9 (3), 20–39.

Hurri, K. (2020). Rethinking climate leadership: Annex I countries' expectations for China's leadership role in the post-Paris UN climate negotiations. *Environmental Development*, 35, 100544.

Hurri, K. (2023a). Climate leadership through storylines: A comparison of developed and emerging countries in the post-Paris era. *Global Society*, 37 (4), 571–592.

Hurri, K. (2023b). The roles they play: Change in China's climate leadership role during the post-Paris era. *Globalizations*, 20 (7), 1065–1082.

Hurri, K., & Kopra, S. (2023). Russia's full-scale invasion of Ukraine: Impacts on China's climate responsibility in the Arctic. November 21. Washington, DC: The Arctic Institute.

Huxham, C. (2003). Action research as a methodology for theory development. *Policy and Politics*, 31 (2), 239–248.

Intergovernmental Panel on Climate Change (IPCC) (2014). *Climate change 2014: Impacts, adaptation, and vulnerability – summary for policy makers*. Working Group II Contribution to the Fifth Assessment Report of the Intergovernmental Panel on Climate Change. Geneva: United Nations.

International Atomic Energy Agency (IAEA) (2024). *Power reactor information system*. Vienna: International Atomic Energy Agency.

International Carbon Action Partnership (ICAP) (2019). *What is emissions trading?* Berlin: International Carbon Action Partnership.

International Energy Agency (IEA) (2020a). Coal 2020. Paris: IEA. Retrieved from https://www.iea.org/reports/coal-2020. (Licence: CC BY 4.0).

International Energy Agency (IEA) (2020b). *Korea 2020: Energy policy review*. Paris: IEA.

International Energy Agency (IEA) (2021a). *An energy sector roadmap to carbon neutrality in China*. Paris: IEA.

International Energy Agency (IEA) (2021b). *Gas market report Q3-2021*. Vienna: IAEA.
International Energy Agency (IEA) (2021c). *Japan 2021: Energy policy review*. Paris: IEA.
International Energy Agency (IEA) (2022a). *An energy sector roadmap to net zero emissions in Indonesia*. Paris: IEA.
International Energy Agency (IEA) (2022b). *U.S.-China joint Glasgow declaration on enhancing climate action in the 2020s*. Vienna: IAEA.
International Energy Agency (IEA) (2023a). *Australia 2023: Energy policy review*. Paris: IEA.
International Energy Agency (IEA) (2023b). *New Zealand 2023: Energy policy review*. Paris: IEA.
International Energy Agency (IEA) (2023c). *Thailand's clean energy transition*. Paris: IEA.
International Monetary Fund (IMF) (2021a). *Coal phase out and energy transition pathways for Asia and the Pacific*. Washington, DC: IMF.
International Monetary Fund (IMF) (2021b). *Beyond zero emissions*. Washington, DC: IMF.
International Monetary Fund (IMF) (2021c). *Fiscal policies to address climate change in Asia and the Pacific*. Washington, DC: IMF.
International Monetary Fund (IMF) (2022). *How replacing coal with renewable energy could pay for itself*. (IMF Blog). Washington, DC: IMF. Retrieved from https://www.imf.org/en/Blogs/Articles/2022/06/08/how-replacing-coal-with-renewable-energy-could-pay-for-itself.
International Panel on Climate Change (IPCC) (2014). Asia. In: *Climate change 2014: Impacts, adaptation, and vulnerability*. Part B: Regional Aspects. [Hijioka, Y., E. Lin, J.J. Pereira, R.T. Corlett, X. Cui, G.E. Insarov, R.D. Lasco, E. Lindgren, and A. Surjan, Contribution of Working Group II to the Fifth Assessment Report of the Intergovernmental Panel on Climate Change [Barros, V.R., C.B. Field, D.J. Dokken, M.D. Mastrandrea, K.J. Mach, T.E. Bilir, M. Chatterjee, K.L. Ebi, Y.O. Estrada, R.C. Genova, B. Girma, E.S. Kissel, A.N. Levy, S. MacCracken, P.R. Mastrandrea, and L.L. White (Contributing Eds.)] (pp. 1327–1370). Cambridge: Cambridge University Press.
International Panel on Climate Change (IPCC) (2022a). *The evidence is clear: The time for action is now. We can halve emissions by 2030*. Media Press release. April 4. Geneva, Switzerland: IPCC. Retrieved from https://www.ipcc.ch/2022/04/04/ipcc-ar6-wgiii-pressrelease/.
International Panel on Climate Change (IPCC) (2022b). *Climate Change 2022: Impacts, adaptation, and vulnerability*. Contribution of Working Group II to the Sixth Assessment Report of the Intergovernmental Panel on Climate Change [H.-O. Pörtner, D.C. Roberts, M. Tignor, E.S. Poloczanska, K. Mintenbeck, A. Alegría, M. Craig, S. Langsdorf, S. Löschke, V. Möller, A. Okem, B. Rama (Contributing Eds.)]. Cambridge: Cambridge University Press.
International Renewable Energy Agency (IRENA) (2022). *NDCs and renewable energy targets in 2021: Are we on the right path to a climate-safe future?* Abu Dhabi: International Renewable Energy Agency.
International Renewable Energy Agency (IRENA) (2023). *Asia and the Pacific: Overview*. Abu Dhabi: IRENA.
International Renewable Energy Agency (IRENA) (2024). *Country rankings*. Vienna: IAEA.
Ishii, K., Macaire, C. & Stalla-Bourdilion, A. (2023). China has reduced its energy bill thanks to Russian oil discounts. Eco Notepad Newsletter. Banque de France. Retrieved from https://www.banque-france.fr/en/publications-and-statistics/publications/china-has-reduced-its-energy-bill-thanks-russian-oil-discounts.

Karlsson, C., Hjerpe, M., Parker, C., & Linner, B.O. (2012). The legitimacy of leadership in international climate change negotiations. *Ambio*, 41 (Suppl 1), 46–55. doi: 10.1007/s13280-011-0240-7.

Kelman, I. (2015). Climate change and the Sendai framework for disaster risk reduction. *International Journal for Disaster Risk Science*, 6, 117–127.

Kemp, J. (2023). Beset by drought, China turned to coal to keep lights on. Reuters Commentary. July 22. Toronto: Reuters.

Keohane, R.O., & Victor, D.G. (2011). The regime complex for climate change. *Perspectives on Politics*, 9 (1), 7–23.

Khan, M.R. (2017). Capacity building for climate change: Lessons from other regimes. SEI papers. Adaptation Watch Brief. Stockholm: Stockholm Environment Institute.

Khan, M.R., Mfitumukiza, D., & Huq, S. (2020). Capacity building for implementation of nationally determined contributions under the Paris Agreement. *Climate Policy*, 20 (4), 499–510.

Konidari, P., & Mavrakis, D. (2007). Multi-criteria evaluation of climate policy instruments. *Energy Policy*, 35 (12), 6235–6257.

Lewis, J.I. (2023). *Cooperating for climate: Learning from international partnerships in China's clean energy sector*. Cambridge: MIT Press.

Li, X. (2021). China: Emerging low-carbon pioneers at city level. In R.K.W. Wurzel, M.S. Anderson & P. Tobin (Eds.), *Climate governance across the globe: Pioneers, leaders, and followers*. (pp. 24–44). London: Routledge.

Lim, C.H., Basu, R., Carrière-Swallow, Y., Kashiwase, K., Kutlukaya, M., Li, M., Refayet, E., Seneviratne, D., Sy, M., & Yang, R. (2024). *Unlocking climate finance in Asia-Pacific: Transitioning to a sustainable future*. Washington, DC: International Monetary Fund.

Liu, H., Evans, S., Zhang, Z., Song, W., & You, X. (2023). *The Carbon Brief profile: China*. London: Carbon Brief. Retrieved from https://interactive.carbonbrief.org/the-carbon-brief-profile-china/.

Mabon, L. (2021). Taiwan's status at the science-policy interface for global climate change: Why getting it right matters [special issue]. Taiwan Research Hub. University of Nottingham.

Maltais, A. (2014). Failing international climate politics and the fairness of going first. *Political Studies*, 62, 618–633.

McGee, J., & Taplin, R. (2008). The Asia-Pacific Partnership and United States international climate change policy. *Colorado Journal of International Environmental Law, and Policy*, 19 (2), 179–218.

Mehling, M., Mielke, S., Fan, C., Hui-Chen, C., & Tsai, W.-C. (2013). *Case study: Taiwan trading carbon across jurisdictions*. Working paper. London: Climate Strategies.

Meidan, M. (2020). China: Climate leader and villain. In M. Hafner & S. Tagliapietra, (Eds.), *The geopolitics of the global energy transition*. (pp. 75–91). *Lecture Notes in Energy*, vol. 73. Cham: Springer.

Ministry of Ecology & Environment of the People's Republic of China (MOE&E) (2022). *China's policies and actions for addressing climate change 2022*. Beijing: MOE&E. Retrieved from https://english.mee.gov.cn/Resources/Reports/reports/202211/P020221110605466439270.pdf.

Ministry of Economy (MOEcon) (2024). *National energy transition roadmap*. Putrajaya, Malaysia: Ministry of Economy.

Ministry of Environment (MOE) (2023). Results of the 24th Tripartite Environment Ministers Meeting among Japan, Korea, and China. Press release November 6. Tokyo: MOE. Retrieved from https://www.env.go.jp/en/press/press_02144.html.

Ministry of Environment (MOE) (2024). Voluntary greenhouse gas reduction offset information platform. Taipei: MOE Climate Change Administration. Retrieved from https://carbonoffset.moenv.gov.tw/EnglishVersion.

Ministry of Foreign Affairs (MOFA) (2022). *Joint statement between the Kingdom of Thailand and the People's Republic of China on working towards a Thailand – China community with a shared future for enhanced stability, prosperity, and sustainability.* Bangkok: Kingdom of Thailand.

Ministry of Foreign Affairs Japan (MOFAJ) (2024). Joint crediting mechanism (JCM). Tokyo: Ministry of Foreign Affairs. Retrieved from https://www.mofa.go.jp/ic/ch/page1we_000105.html.

Muta, T., & Erdogan, M. (2023). The global energy crisis pushed fossil fuel consumption subsidies to an all-time high in 2022. International Renewable Energy Agency News commentary, February 16. Vienna: IAEA.

Myllyvirta, L., & Qin, Q. (2023). China's carbon dioxide (CO2) emissions grew 10% year-on-year in the second quarter of 2023, rising approximately 1% above the record levels seen in 2021. August 10. London: Carbon Brief.

Nakamura, A., & Crowley, K. (2016). Addressing the resilience-community dynamic: Climate change adaptation policy, natural disaster management and the policy cycle. Aichi-Nagoya, 36th Annual Conference of the International Association for Impact Assessment [IAIA], May 11–14.

Nakamura, A., & Crowley, K. (2020). Regional action and carbon trading in the East Asia Pacific. In H.V. Hanh & P.T. Song Thuong (Eds.), *Sustainable regional development: Theoretical and practical issues.* (pp. 519–550). Ha Noi, Viet: Social Sciences Publishing House.

Nangoy, F., & Vu, K. (2023). Indonesia, Viet Nam energy transition finance under JETP. *Reuters News.* September 25.

Napoli, C. (2012). Understanding Kyoto's failure. *The SAIS Review of International Affairs*, 32 (2), 183–196.

New Zealand Foreign Affairs and Trade. (NZFA&T) (2023). China. News. Wellington, NZ: NZFA&T. Retrieved from https://www.mfat.govt.nz/en/countries-and-regions/asia/china/#.

Notre Dame Global Adaptation Initiative (ND-GAIN) (2021). *ND-GAIN country index rankings.* Notre Dame: University of Notre Dame.

Notre Dame Global Adaptation Initiative (ND-GAIN) (2022). *Helping countries and cities counter the risks of a changing climate.* Notre Dame: University of Notre Dame.

Notre Dame Global Adaptation Initiative (ND-GAIN) (2023). *Country Index.* Notre Dame USA: University of Notre Dame.

Nyonho, O., Miteva, D.A., & Lee, Y. (2023). Impact of Korea's emissions trading scheme on publicly traded firms. *Plos One*, 18 (5), e0285863.

Organisation for Economic Co-operation and Development (OECD) (2017). Designing carbon pricing instruments for ambitious climate policy: Discussion paper. Carbon Market Platform 2nd Strategic Dialogue Rome, September 27–28. Paris: OECD publishing.

Organisation for Economic Co-operation and Development (OECD) (2022a). *Aggregate trends of climate finance provided and mobilised by developed countries in 2013-2020: Climate finance and the USD 100 billion goal.* Paris: OECD.

Organisation for Economic Co-operation and Development (OECD) (2022b). *Pricing greenhouse gas emissions: Turning climate targets into climate action.* Paris: OECD.

Organisation for Economic Co-operation and Development & International Energy Agency (OECD & IEA) (2015). Energy and climate change. World Energy Outlook Special Report. Paris: IEA.

Ostrom, E. (2010). Polycentric systems for coping with collective action and global environmental change. *Global Environmental Change*, 20, 550–557.

Ourbak, T., & Magnan, A.K. (2018). The Paris Agreement and climate change negotiations: Small islands, big players. *Regional Environmental Change*, 18, 2201–2207.

Overseas Development Institute & Heinrich Böll Stiftung (ODI & HBS) (2022). *The principles and criteria of public climate finance: A normative framework.* Washington, DC: HBS.

Oxford Committee for Famine Relief (Oxfam) (2022). *Climate finance short-changed: The real value of the $100 billion commitment in 2019–20.* Nairobi: Oxfam.

Pacific Community. (2022). Can our region be carbon neutral by 2050? Press Release, August 24. Suva: Pacific Community. Retrieved from https://www.spc.int/updates/blog/2022/08/can-our-region-be-carbon-neutral-by-2050.

Pacific Community. (2016). *Framework for resilient development in the Pacific: An integrated approach to address climate change and disaster risk management 2017-2030 (FRDP).* Noumea: Pacific Community.

Pacific Economic Cooperation Council (PECC) (2021a). Pacific Economic Cooperation Council member committees. Singapore: PECC. Retrieved from https://www.pecc.org/about/member-committees.

Pacific Economic Cooperation Council (PECC) (2021b). *State of the region 2021-2022.* Singapore: PECC. Retrieved from https://www.pecc.org/resources/regional-cooperation/2703-state-of-the-region-2021-2022.

Pacific Islands Forum (PIF) (2019). *State of Pacific regionalisation.* Suva: PIF. Available via https://catalogue.nla.gov.au/catalog/881268 (previously viewed at https://www.forumsec.org/wp-content/uploads/2019/10/State-of-regional-Report-2019-1.pdf).

Pacific Islands Forum (PIF) (2022). *The 2050 strategy for the Blue Pacific Continent.* Suva: PIF.

Pacific Resilience Partnership (PRP) (2023). Pacific resilience partnership members. Suva: PIF.

Pauw, P., Mbeva, K., & Asselt, V.H. (2019). Subtle differentiation of countries 'responsibilities' under the Paris Agreement. *Palgrave Communications*, 5(1). Online publication. Retrieved from https://ideas.repec.org/a/pal/palcom/v5y2019i1d10.1057_s41599-019-0298-6.html.

Pearse, R., & Böhm, S. (2014). Ten reasons why carbon markets will not bring about radical emissions reduction. *Carbon Management*, 5 (4), 325–337.

People's Republic of China (PRC) (2023a). Joint statement between the People's Republic of China and New Zealand on the comprehensive strategic partnership. June 29. Beijing: The State Council.

People's Republic of China (PRC) (2023b). Han Zheng meets with Prime Minister of Malaysia Anwar Ibrahim. Press release. November 9. Beijing: Ministry of Foreign Affairs.

People's Republic of China in the Federated States of Micronesia embassy (PRC/FSM) (2024). *Fact Sheet: Cooperation between China and Pacific Island countries.* Pohnpei, Micronesia: PRC/FSM. Retrieved from http://fm.china-embassy.gov.cn/eng/zgxw/202205/t20220524_10691917.htm.

Pike, L. (2019). 'Green Belt and Road' in the spotlight. China Dialogue, April 24. London: China Dialogue Trust. Retrieved from https://chinadialogue.net/en/energy/11212-green-belt-and-road-in-the-spotlight/

Powles, A., & Wallis, J. (2022). Can Pacific Islands forum learn anything from ASEAN? *East Asia Forum*, August 14.

Price Waterhouse & Cooper. (PwC) (2023a). *Code red: Asia Pacific's time to go green.* London: PwC.

Price Waterhouse & Cooper. (PwC) (2023b). *Viet Nam's eighth power development plan: Insights and considerations for investors.* Hanoi and Ho Chi Minh City: PwC.

Prime Minister of Australia (PMA) (2023). Statement on joint outcomes of the China-Australia annual leaders' meeting. Media statement. November 7. Canberra: PMA.

Psaropoulos, J. (2023). Ukraine war sped the world on a path to net zero emissions. Climate features. Media Report. Doha: Al-Jazeera, September 26.

Qi, J., & Dauvergne, P. (2022). China's rising influence on climate governance: Forging a path for the global south. *Global Environmental Change*, 73, 102484.

Reid, R., & Botterill, L.C. (2013). The multiple meanings of 'resilience': An overview of the literature. *Australian Journal of Public Administration*, 72 (1), 31–40.

Reklev, S. (2022). South Korea lines up potential partners for Article 6.2 carbon trade. *Carbon Pulse*, August 23.

ROC Taiwan. (2023). Public, private Taiwanese actors participate in UNFCCC COP28, highlight Taiwan's contribution to a net-zero world. Press release. December 19. Taipei: Ministry of Foreign Affairs, Republic of China (Taiwan).

Rogelj, J., Shindell, D., Jiang, K., Fifita, S., Forster, P., Ginzburg, V., Handa, C., Kheshgi, H., Kobayashi, S., Kriegler, E., Mundaca, L., Séférian, R., & Vilariño, M.V. (2018). Mitigation pathways compatible with 1.5°C in the context of sustainable development. In V. Masson-Delmotte, P. Zhai, H.-O. Pörtner, D. Roberts, J. Skea, P.R. Shukla, A. Pirani, W. Moufouma-Okia, C. Péan, R. Pidcock, S. Connors, J.B.R. Matthews, Y. Chen, X. Zhou, M.I. Gomis, E. Lonnoy, T. Maycock, M. Tignor & T. Waterfield (Eds.), *Global warming of 1.5°C: An IPCC Special Report on the impacts of global warming of 1.5°C above pre-industrial levels and related global greenhouse gas emission pathways, in the context of strengthening the global response to the threat of climate change, sustainable development, and efforts to eradicate poverty.* (pp. 93–174). Cambridge, UK and New York, NY: Cambridge University Press.

Romsom, E., & McPhail, K. (2020). *The energy transition in Asia: Country priorities, fuel types, and energy decisions.* WIDER Working Paper 2020/48. United Nations University World Institute for Development Economics Research.

Ronaghi, M., & Scorsone, E. (2023). The impact of COVID-19 outbreak on CO_2 emissions in the ten countries with the highest carbon dioxide emissions. *Journal of Environmental Public Health*, June 13, 4605206.

Rosales, R. (2023). How ASEAN countries are preparing for carbon market growth. *Sustainable Views.* March 8.

Sandalow, D., Meidan, M., Andrews-Speed, P., Hore, A., Qui, S., & Downie, E. (2022). *Guide to Chinese climate policy.* Oxford: Oxford Institute for Energy Studies.

Savage, L. (2019). The US left a hole in leadership on climate. China is filling it. Arlington: VA: Politico. August 15. Online opinion. Retrieved from https://www.politico.com/story/2019/08/15/climate-china-global-translations-1662345.

Schipper, L., & Pelling, M. (2006). Disaster risk, climate challenge and international development: Scope for, and challenges to, integration. *Disasters*, 30 (1), 19–38.

Schive, C. (1990). The next stage of industrialization in Taiwan and South Korea. In G. Gereffi & D. Wyman (Eds.), *Manufacturing miracles: Paths of industrialization in Latin America and East Asia.* (pp. 267–291). Princeton, NJ: Princeton University Press.

Schreurs, M.A. (2017). Multi-level climate governance in China. *Environmental Policy Governance*, 27 (2), 163–174.

Schreurs, M.A., & Tiberghien, Y. (2007). Multi-level reinforcement: Explaining European Union leadership in climate change mitigation. *Global Environmental Politics*, 7 (4), 19–46.

Schwerhoff, G., & Kornek, U. (2018). Leadership in climate change mitigation: Consequences and incentives. *Journal of Economic Surveys*, 32 (2), 491–517.

Seah, S., & Martinus, M. (2021). *Gaps and opportunities in ASEAN's climate governance.* Singapore: ISEAS Publishing.

Shaw, A. (2024). The future of coal: Why China's appetite remains. *Mining Technology.* London and New York. February 22.

Sloan, J. (2021). Fiji's Climate Change Act, 2021 – Fiji's whole of government approach to reduce emissions and remove carbon. *Ocean Law Bulletin.* September 29.

Soezer, A. (2022). What is Article 6 of the Paris Agreement and why is it important? UNDP *Sustainable Energy Hub news.* November 9.

Statista. (2023a). Emissions. Hamburg: Statistica. Retrieved from https://www.statista.com/markets/408/topic/949/emissions/#overview.

Statista. (2023b). Overseas energy dependence ratio South Korea 2011-2021. Hamburg: Statista. Retrieved from https://www.statista.com/statistics/1370949/south-korea-overseas-energy-dependence-ratio/.

Stewart, R.B., Oppenheimer, M., & Rudyk, B. (2013). Building blocks for global climate protection. *Stanford Environment Law Review*, 32 (2), 341–392.

Stokes, L.C. & Mildenberger, M. (2020). The trouble with carbon pricing: Only a bold approach that centres politics can meet the scale of the climate crisis. *Boston Review*. September 24.

Sun, X., & Mi, Z. (2023). Factors driving China's carbon emissions after the COVID-19 outbreak. *Environmental Science and Technology*, 57 (48), 19125–19136.

Suroyo, G. & Sulaiman, S. (2023). Indonesia to seek China's help to develop renewables at Belt and Road Forum. *Reuters Daily Briefing News*. October 15.

Tan, C. (2023). Singapore sets eligibility criteria for carbon credits that can be used to offset carbon tax. *The Straits Times*. October 13.

The Pacific Community (SPC) (2022). Taking action on climate change to shape a resilient Pacific. Noumea: SPC. Retrieved from https://www.spc.int/fr/documentation/publications/taking-action-on-climate-to-shape-a-resilient-pacific.

Thomas, A., Baptiste, A., Martyr-Koller, R., Pringle, P., & Rhiney, K. (2020). Climate change and small island developing states. *Annual Review of Environment and Resources*, 45, 1–27.

Thomas, V., Albert, J.R.G., & Perez, R.T. (2013). *Climate-related disasters in Asia and the Pacific*. ADB Economics Working Paper No. 358. Manila: Asian Development Bank.

Tobin, P. (2017). Leaders and laggards: Climate policy ambition in developed states. *Global Environmental Politics*, 17 (4), 28–47.

Tobin, P., Wurzel, R.K.W., & Andersen, M.S. (2021). Conclusion: Pioneers, leaders and followers in multilevel and polycentric climate governance reassessed. In R.K.W. Wurzel, M.S. Anderson & P. Tobin (Eds.), *Climate governance across the globe: Pioneers, leaders, and followers*. (pp. 259–276). London: Routledge.

Torvanger, A., & Ringius, L. (2002). Criteria for evaluation of burden-sharing rules in international climate policy. *International Environmental Agreements*, 2 (3), 221–235.

Twidale, S. (2023). Global carbon pricing schemes raised record $95 bln in 2022 – World Bank. *Reuters*. May 23. Retrieved from https://www.reuters.com/markets/commodities/global-carbon-pricing-schemes-raised-record-95-bln-2022-world-bank-2023-05-23/.

Underdal, A. (1994). Leadership theory: Rediscovering the arts of management. In I.W. Zartman (Ed.), *International multilateral negotiation: Approaches to the management of complexity*. (pp. 178–197). San Francisco: Jossey-Bass Publishers.

United Nations (UN) (2021). Secretary-General calls latest IPCC climate report 'code red for humanity', stressing 'irrefutable' evidence of human influence. Press release, Secretary General. New York: United Nations.

United Nations (UN) (2022a). Secretary-General warns of climate emergency, calling Intergovernmental Panel's report 'a file of shame', while saying leaders 'are lying', fuelling flames. Press release, Secretary General. New York: United Nations.

United Nations (UN) (2022b). COP27 reaches breakthrough agreement on new "loss and damage" fund for vulnerable countries. Press release. November 20. Bonn: UNFCCC.

United Nations (UN) (2023a). *Introduction to climate finance*. New York: United Nations. Retrieved from https://unfccc.int/topics/introduction-to-climate-finance.

United Nations (UN) (2023b). Hottest July ever signals 'era of global boiling has arrived' says UN chief. Press release, Secretary General. July 27. New York: United Nations.

United Nations Climate Change Secretariat (UNCCS) (2017). *Opportunities and options for integrating climate change adaptation with the Sustainable Development Goals and the Sendai Framework for Disaster Risk Reduction 2015–2030.* Bonn, Germany: UNCCS.
United Nations Department of Economic and Social Affairs (UNESA) (2007). Small Island Developing States. New York: UNESA.
United Nations Development Programme (UNDP) (2021). *Nationally determined contributions (NDC) global outlook report 2021: State of climate ambition.* New York: UNDP.
United Nations Economic and Social Commission for Asia and the Pacific (UNESCAP) (2020). *The impact and policy responses for COVID-19 in Asia and the Pacific.* Bangkok: UNESCAP.
United Nations Economic and Social Commission for Asia and the Pacific (UNESCAP) (2022). *Asia-Pacific 'riskscapes' @1.5C: Sub-regional pathways for adaptation and resilience.* Bangkok: UNESCAP.
United Nations Economic and Social Commission for Asia and the Pacific (UNESCAP) (2023a). *Review of Climate Ambition in Asia and the Pacific: Just transition towards regional net-zero climate resilient development.* Bangkok: UNESCAP.
United Nations Economic and Social Commission for Asia and the Pacific (UNESCAP) (2023b). About ESCAP – Our work. Bangkok: UNESCAP.
United Nations Economic and Social Commission for Asia and the Pacific (UNESCAP) (2023c). *The race to net zero: Accelerating climate action in Asia and the Pacific.* Bangkok: UNESCAP.
United Nations Economic and Social Commission for Asia and the Pacific (UNESCAP) (2023d). *Building (Asia and the Pacific) resilience to cascading risks, including disasters, climate change and health crises.* Bangkok: UNESCAP.
United Nations Environment Programme (UNEP) (2022). *Adaptation Gap Report. Too little. Too slow: Climate adaptation failure puts world at risk.* Nairobi: UNEP.
United Nations Environment Programme (UNEP) (2023). United Nations Environment Programme. *Emissions gap report 2021: Temperatures hit new highs, yet the world fails to cut emissions (again).* Nairobi: UNEP.
United Nations Framework Convention on Climate Change (UNFCCC) (1992). *United Nations Framework Convention on Climate Change.* Bonn: Climate Change Secretariat, UNFCCC.
United Nations Framework Convention on Climate Change (UNFCCC) (2005). *Climate change, small island developing states.* Bonn: Climate Change Secretariat, UNFCCC.
United Nations Framework Convention on Climate Change (UNFCCC) (2007). *Report of the Conference of the Parties serving as the meeting of the Parties to the Kyoto Protocol.* Geneva: United Nations.
United Nations Framework Convention on Climate Change (UNFCCC) (2015). *Paris Agreement.* Bonn: Climate Change Secretariat, UNFCCC.
United Nations Framework Convention on Climate Change (UNFCCC) (2019). *Climate action and support trends.* Bonn: Climate Change Secretariat, UNFCCC.
United Nations Framework Convention on Climate Change (UNFCCC) (2020). *Defining the 'Starting Line.' UNFCCC Race to Zero Campaign.* Bonn: Climate Change Secretariat, UNFCCC.
United Nations Framework Convention on Climate Change (UNFCCC) (2022). Greenhouse gas emissions – Time Series – Annex 1. Bonn: Climate Change Secretariat, UNFCCC. Retrieved from https://di.unfccc.int/time_series.
United Nations Framework Convention on Climate Change (UNFCCC) (2023). *About carbon pricing.* Bonn: Climate Change Secretariat, UNFCCC.
United Nations Framework Convention on Climate Change (UNFCCC). (2024a). NDC registry. Bonn: UNFCCCC. Retrieved from https://unfccc.int/NDCREG.

United Nations Framework Convention on Climate Change (UNFCCC). (2024b). Pledges to the loss and damage fund. Bonn: UNFCCCC. Retrieved from https://unfccc.int/process-and-meetings/bodies/funds-and-financial-entities/loss-and-damage-fund-joint-interim-secretariat/pledges-to-the-loss-and-damage-fund.

United Nations International Children's Emergency Fund (UNICEF) (2023). *UNICEF East Asia and Pacific Region humanitarian situation report No. 1* (January 1 to 30 June 30). New York: UNICEF.

United Nations Office for Disaster Risk Reduction (UNDRR) (2021). *Asia-Pacific action plan 2021-2024 for the implementation of the Sendai Framework for Disaster Risk Reduction 2015-2030.* Brussels: United Nations Office for Disaster Risk Reduction – Regional Office for Europe & Central Asia.

United Nations Office for Disaster Risk Reduction (UNDRR) (2024). *Strengthening disaster resilience and accelerating the implementation of Sendai Framework for Disaster Risk Reduction in Central Asia.* Brussels: United Nations Office for Disaster Risk Reduction – Regional Office for Europe & Central Asia.

United Nations, Economic and Social Commission for Asia and the Pacific, United Nations Environment Programme, International Labour Organization, the Regional Collaboration Centre for Asia-Pacific of the Secretariat of the United Nations Framework Convention on Climate Change, United Nations Industrial Development Organization (UNESCAP, ILO, UNFCCC/RCC Asia-Pacific & UNIDO) (2023a). *Review of climate ambition in Asia and the Pacific: Just transition towards regional net-zero climate resilient development.* Bangkok: United Nations.

Uren, D. (2023). China's Belt and Road initiative and quasi-IMF lending. The Strategist, October 19. Canberra: Australian Strategic Policy Institute.

Verstegen, P. (2023). *Safeguarding fossil fuels: How loopholes in the Australian government's climate policy will safeguard a future for fossil fuels at the expense of Australian businesses, consumers, and the climate.* Melbourne: Australian Conservation Foundation.

Vietnam Law and Legal Forum (VLLF) (2023). Vietnam – China joint statement. *News – Vietnam Law and Legal Forum.* December 14. Retrieved from https://vietnamlawmagazine.vn/vietnam-china-joint-statement-70965.html.

Wang, C. (2024). *China Belt and Road initiative (BRI) investment report 2023.* Shanghai: Green Finance & Development Centre, Fudan University. Retrieved from https://greenfdc.org/china-belt-and-road-initiative-bri-investment-report-2023/.

Woetzel, J., & Sengupta, J. (2022). Asia Pacific should use its wealth more productively. *Nikkei Asia.* January 30.

World Bank (2021). *Climate change action plan (2021-2025): Supporting green, resilient, and inclusive development.* Washington, DC: World Bank.

World Bank (2022a). *World Bank in East Asia Pacific: Overview.* Washington, DC: World Bank.

World Bank. (2022b). 10 things you should know about the World Bank Group's climate finance. Washington, DC: World Bank. Retrieved from https://www.worldbank.org/en/news/factsheet/2022/09/30/10-things-you-should-know-about-the-world-bank-group-s-climate-finance.

World Bank (2022c). China: Country climate and development report. Washington, DC: World Bank. Retrieved from https://openknowledge.worldbank.org/server/api/core/bitstreams/35ea9337-dfcf-5d60-9806-65913459d928/content.

World Bank (2023a). *Climate and development in East Asia and Pacific region.* Washington, DC: World Bank.

World Bank (2023b). *Turn up the heat: Climate extremes, regional impacts, and the case for resilience.* Washington, DC: World Bank.

World Bank (2023c). *State and trends of carbon pricing 2023.* Washington, DC: World Bank.

World Bank (2023d). Carbon pricing dashboard. Washington, DC: World Bank.

World Bank (2024a). *Climate action in East Asia and Pacific: Urgent challenges, innovative solutions.* Washington, DC: World Bank.
World Bank. (2024b). World bank open data. Washington, DC: World Bank. Retrieved from https://data.worldbank.org/.
World Health Organization (WHO) (2023). *Finance for health and climate change.* Geneva: WHO.
World Meteorological Organisation (WMO). (2021). *State of the climate in Asia 2020.* Geneva: WMO.
World Meteorological Organisation (WMO) (2023). *State of the climate in Asia 2022.* Geneva: WMO.
World Meteorological Organisation (WMO). (2024). *State of the climate in Asia 2023.* Geneva: WMO.
World Population Review (WPR) (2023). APAC countries – Asia-Pacific countries 2023. California: WPR. Retrieved from https://worldpopulationreview.com/country-rankings/apac-countries.
World Population Review (WPR) (2024). Solar power by country 2024. Walnut, California: WPR. Retrieved from https://worldpopulationreview.com/country-rankings/solar-power-by-country.
World Resources Institute (WRI) (2023). World's top 10 emitters, and how they've changed. Washington, DC: WRI. Retrieved from http://www.wri.org/blog/2017/04/interactive-chart-explains-worlds-top-10-emitters-and-how-theyve-changed.
Wu, S. (2023). A systematic review of climate policies in China: Evolution, effectiveness, and challenges. *Environmental Impact Assessment Review*, 99, 107030.
Wurzel, R.K.W., Anderson, M.S., & Tobin, P. (2021). Climate governance across the globe: Pioneers, leaders, and followers. In R.K.W. Wurzel, M.S. Anderson & P. Tobin (Eds.), *Climate governance across the globe: Pioneers, leaders, and followers.* (pp. 3–20). London: Routledge.
Wurzel, R.K.W., Andersen, M.S. & Tobin, P. (Eds.) (2021). *Climate governance across the globe: Pioneers, leaders, and followers.* New York: Routledge.
Wurzel, R.K.W., Liefferink, D., & Torney, D. (2019). Pioneers, leaders, and followers in multilevel and polycentric climate governance. *Environmental Politics*, 28 (1), 1–21.
Yang, J. (2022). Understanding China's challenging engagement in global climate governance: A struggle for identity. *Asia Europe Journal*, 20, 357–376.
Yau, R., & Chen, G.H. (2021). Assessing energy subsidy policies in a structural macroeconomic model. *Energy Economics*, 103, 105509.
Zhang, D. (2022). China's influence as a Pacific donor. *The Interpreter – Lowy Institute.* October 31.
Zhang, Z.X. (2023). From China's climate commitments to action. East Asia Forum, March 25. Canberra: Australian National University.

Index

Note: *Italicized* and **bold** pages refer to figures and tables, respectively.

FSC
www.fsc.org
MIX
Papier | Fördert gute Waldnutzung
FSC® C083411

Kolibri 360 GmbH
Ferdinand-Jühlke-Straße 7
99095 Erfurt, Deutschland
produktsicherheit@kolibri360.de